Elizabeth Siegel Watkins

On the Pill

A Social History of
Oral Contraceptives
1950–1970

The Johns Hopkins University Press
BALTIMORE AND LONDON

© 1998   The Johns Hopkins University Press
All rights reserved. Published 1998
Printed in the United States of America
on acid-free paper
9 8 7 6 5 4 3 2 1

The Johns Hopkins University Press
2715 North Charles Street
Baltimore, Maryland 21218–4363
The Johns Hopkins Press Ltd., London
w w w . p r e s s . j h u . e d u

Library of Congress Cataloging-in-Publication Data
will be found at the end of this book.
A catalog record for this book is available from the
British Library.

ISBN 0-8018-5876-3

Contents

Acknowledgements   vii

        Introduction   1
One   Genesis of the Pill   9
Two   Physicians, Patients, and the New Oral Contraceptives   34
Three   Sex, Population, and the Pill   53
Four   Debating the Safety of the Pill   73
Five   Oral Contraceptives and Informed Consent   103
Six   Conclusion   132

        Notes   139
        Bibliographical Essay   163
        Index   175

Illustrations follow p. 72

Acknowledgments

Although at times I felt very much alone when writing this book, I received help along the way from many individuals and organizations. I owe a debt of gratitude to the following institutions for financial support: Harvard University, the National Science Foundation, the Woodrow Wilson National Fellowship Foundation, and the Rockefeller Archive Center. I am also very grateful to the individuals who granted me interviews: Ben Gordon, Morton Mintz, Judy Norsigian, Barbara Seaman, and Alice Wolfson.

I was ably assisted in my research by the staff members of the following libraries and archives: Schlesinger Library at Radcliffe College, the Sophia Smith Collection at Smith College, the Library of Congress, the American College of Obstetrics and Gynecology, Countway Library of Medicine at Harvard University, the Boston Women's Health Book Collective, and the Worcester Foundation for Experimental Biology. I would like to extend special thanks to Patricia Gossel at the National Museum of American History, Suzanne White Junod at the Food and Drug Administration Historian's Office, and Tom Rosenbaum at the Rockefeller Archive Center, all of whom went out of their way to find the sources I needed.

Everett Mendelsohn had confidence in me from beginning to end; I thank him for his advocacy, encouragement, and guidance. Many individuals contributed thoughtful comments, questions, and suggestions, bringing my ideas into sharper focus. Thanks to Allan Brandt, Barbara Rosenkrantz, Anne Harrington, Garland Allen, Dorothy Porter, Betsy Smith, Mark Barrow, David Spanagel, and Simon Watkins for the parts they played in the conceptualization, research, and writing of this book.

Several more people helped to transform the manuscript into a book. Lara Marks read the manuscript carefully, and her insightful review helped me to revise and refine my work. The copyediting skills of Ruth Haas rescued my writing from tendencies toward repetitive and awkward prose. It was a pleasure to work with the staff at Johns Hopkins University Press, and I am especially grateful to

Bob Brugger for putting his faith in this book and for offering critical editorial guidance, and to Carol Zimmerman for smoothly facilitating the book's production.

I give heartfelt thanks to my parents, my brother, and my friends for their love, support, diversion, and child care, and to my daughters, Emily and Ellen, for understanding when I had to work and for giving me something else to do the rest of the time. Finally, one person has been an integral part of this project from the very first conversation in the car traveling down Interstate 95 one November evening in 1991 to the appearance of this book on bookstore shelves almost seven years later. My husband Simon has been my most steadfast supporter and my best friend, encouraging me when I felt like quitting, making me laugh when I felt like crying, leaving me alone when I wanted space, and keeping me company when I needed it most. It is with love, affection, and eternal gratitude that I dedicate this book to him.

On the Pill

Introduction

In 1968, a popular writer ranked the pill's importance with the discovery of fire and the developments of tool-making, hunting, agriculture, urbanism, scientific medicine, and nuclear energy.[1] Twenty-five years later, the leading British weekly, the *Economist*, listed the pill as one of the seven wonders of the modern world.[2] The image of the oral contraceptive as revolutionary persists in popular culture, yet the nature of the changes it supposedly brought about has not been fully investigated. After more than thirty-five years on the market, the role of the pill is due for a thorough examination.

This book evaluates commonly held assumptions about the impact of the pill on middle-class American society and explores the changing perceptions of the pill in the context of the 1950s and 1960s. The birth control pill had far different consequences than its developers intended. Taken twenty days each month, it altered the female hormonal cycle, preventing ovulation and pregnancy. Scientists and birth control advocates conceived of the pill as a scientifically based, technological solution to the social problems of family planning and population control. Its rapid acceptance as the preferred method of birth control among both women and doctors in the first half of the 1960s far exceeded anyone's expectations. Yet, later in the decade, this early popularity clashed with publicity on the pill's adverse health effects, producing a forceful feminist critique of the pill and, more broadly, of the male medical profession and its role in women's reproductive health. Along the way, the pill generated great interest among the medical profession, the pharmaceutical industry, the federal government, family planning organizations, feminist groups, the media, and the public, and the missions of these groups sometimes conflicted as they interpreted the meaning of the pill differently. Their interpretations clashed notably in regard to three spheres of social activity during the 1960s: the liberalization of sexual attitudes and practices, the medicalization of birth control, and the rise of the new feminism. The history of the pill during the turbulent decade of the sixties reflected changing conceptions of sex, medicine, and technology in American society.

One of the most enduring assumptions about the oral contraceptive credits, or blames, the pill for giving rise to the sexual revolution of the 1960s. Contemporary commentators proclaimed that the pill encouraged the loosening of sexual attitudes and behavior during the turbulence of that decade. The image of the pill as revolutionary took hold and persisted through the 1970s and 1980s to the present. In 1984, a well-known scientist wrote that "the fact that contraception has become an accepted form of dinner conversation . . . is directly associated with the Pill, as is much of the sexual revolution that occurred in the '60s and '70s."[3]

The pill certainly ushered in a contraceptive revolution when it came onto the market in 1960, but the contraceptive revolution did not cause the sexual revolution. In fact, no data have ever supported such an association. In the 1960s and early 1970s, demographers focused on the contraceptive habits of *married* women to document the *contraceptive* revolution, while sociologists surveyed the sexual attitudes and practices of *unmarried* women to study the *sexual* revolution. Journalists combined the two contemporaneous changes and developed the lasting image of the pill as symbol of the sexual revolution; scientists and the public accepted and promoted this interpretation of the pill.

An evaluation of the inconsistent meanings and perceptions of the pill during the 1960s provides insights into the social and cultural attitudes toward sex and sexuality at that time. In the mid-1960s, the use of the pill by an unmarried woman was judged differently, depending on her social class. Many people frowned upon single women from the middle or upper classes using the pill, or any form of birth control, because it implied, correctly, not only that these women were having sex but also that they were planning ahead for it. On the other hand, the moral implications of sexual activity among lower-class unmarried women elicited less concern than did the economic effect of unwanted babies. Demographers in the 1960s successfully focused the public's attention on the potential "crisis" of overpopulation; in that climate, advocates of population control urged widespread use of birth control, including the pill, to slow population growth in the United States and abroad. Later in the decade, discussion about the use and impact of the pill based on moral and political considerations gave way to medical concerns about its safety. By the late 1960s, health issues came to dominate public discourse on the pill.

Any examination of the influence of the pill must include its impact on medical practice and the doctor-patient relationship.[4] Concern about the adverse health effects of the pill compelled both doctors and patients to reconsider risk in choosing a birth control method. Women's contraceptive choices in the late 1960s involved a difficult risk-benefit calculation: the benefit of a highly effective contraceptive versus the risk of potentially fatal complications, or the benefit of a barrier contraceptive with no side effects (e.g., the diaphragm, the condom) versus a higher risk of pregnancy. The absence of safe, legal abortion as a reliable, ultimate backup measure further complicated the risk assessment. Women wanted more information so that they could decide whether to use oral contraceptives, but in light of inconclusive scientific evidence, doctors struggled with what or how much to tell them. Women's requests for information and their physicians' inability or reluctance to provide adequate information strained relations between women patients and doctors and by 1970 increased the distance between consumers and providers of health care.

Oral contraceptives also played a role in the increasing "medicalization" of women's health care and the growing critique of medicine in the late 1960s. The feminist critique of medicine grew from several sources of dissatisfaction, but the controversy over the safety of the pill and the importance of informed consent in its use served as a catalyst for the growth of the women's health movement. In the early 1960s, women who requested oral contraceptives from their physicians became more active participants in their medical care and in so doing ultimately helped to shift the balance of power in the traditional doctor-patient relationship. Later in the decade, as the safety of the pill came into question, women felt confident enough to doubt their physicians' judgment and to demand full disclosure so that they could make their own informed decisions about whether to take the pill.

The debate over informed consent and oral contraceptives also had the effect of increasing government regulation in the practice of medicine. Although the government had become involved in some areas, such as the licensing of physicians, funding of medical research, and regulation of drugs, physicians had successfully blocked government intervention in the doctor-patient relationship before 1970. That year, in response to the growing public controversy over the safety of the pill, the U.S. Senate Subcommittee on Monopoly of the Select Committee on Small Business held a series of hearings on oral contraceptives as part of its investigation of the drug industry. One outcome was a directive from the Food and Drug Administration (FDA) requiring doctors to provide their female patients with information on the potential health hazards associated with oral contraception. The debate over the safety of the pill and government regulation of the interactions between doctors and patients when prescribing oral contraceptives presented a new challenge to the authority of physicians.

The ways in which different groups responded to these issues reveal conflicting interests in the pill in particular and in birth control and women's health care in general. New voices, such as those within the rising feminist movement, presented interpretations of the pill that were not anticipated by its developers and advocates of the previous generation. Although the story of the pill continues to the present, this study ends after the pill's first decade, in the watershed year of 1970, when the FDA ordered manufacturers to include an informational pamphlet on the health risks of oral contraceptives in every package of birth control pills. The lengthy inserts in tiny type found in many prescription drug packages today are the legacy of that mandate.

The Pill and Middle-Class American Women

The story of the pill played out differently in countries around the world. This book explores the meaning of the pill as a new tool of birth control in the United States and reflects on the special contexts of American society at midcentury. The issue of the export of contraceptives to encourage population control programs in the Third World is an important one, but is best framed in a different context, and therefore is not treated here.[5] It is important to distinguish between "birth control" and "population control."[6] The former describes the attempts of individual women to plan when and how many children to have. The latter denotes efforts to control fertility (i.e. birth rate), usually in lower socioeconomic groups or in countries considered less developed. Although they both employ contraceptive methods, birth control and population control seek to achieve different goals.

This analysis concentrates mainly on white, middle-class women, who constituted the largest group of pill users in the United States during the period discussed in this book. This is not to say that low-income women or women of color did not use the pill; many of the statistics and stories described here apply to women regardless of race, ethnicity, or income level. However, not all women shared the same experiences with the pill in particular or with birth control and health services in general. Because the pill was only available by prescription, women had to have the money, time, and means to visit either a private physician or a family planning clinic. That hurdle denied many women access to oral contraceptives. The experiences of those who did obtain the pill may have been different, shaped according to the circumstances under which they sought birth control. The consequences of the pill discussed in this work pertain primarily to white middle- and upper middle-class American society. They encompass the experiences of millions of American women whose lives were affected in some way by the development of the birth control pill, as well as the institutional effects felt by the medical profession, the pharmaceutical industry, and family planning organizations.

Interpretations of the Pill

To understand the social and scientific motivations that led to the pill, I begin by exploring the context in which the oral contraceptive was developed. Several retrospective accounts of the development of the pill written by scientist-participants and by journalists have presented its history as a neat path of scientific advances leading inevitably to the production of the first birth control pill.[7] Knowing the nature of the finished product, namely, the synthetic steroidal composition of the oral contraceptive, these historians used hindsight to recognize the predecessors of the modern-day pill. This study surveys the state of knowledge in relevant areas of inquiry to ascertain what was available to the scientists involved in the development of an oral hormonal contraceptive, and places their research within both the developing intellectual discipline and the larger social and economic context.

Early historical analyses of the pill also reflected the social climate in which they were written. Feminists in the 1950s extolled the birth control pill as a scientific triumph for women in their efforts to gain control over their reproductive lives; the next generation of feminists interpreted the pill quite differently. Starting in the 1970s, feminist scholars articulated a new critique of the birth control pill. The radical version of this argument portrayed the pill as an ill-conceived, poorly tested contraceptive foisted on women through the collusion of the drug industry and the medical profession. The problem with these claims is that they rested upon assumptions about the 1950s and early 1960s rooted in the logic and social politics of the 1970s. When Gena Corea wrote in 1977, "in developing contraceptives, male physicians and researchers have devalued women," her retrospective interpretation of scientists' motivations and women's needs in the pre-pill era did not consider the very different social climate of the 1950s nor the role that feminist leaders of the 1950s played in the pill's development.[8]

Feminists of the next decade began to move away from the narrow indictment of men. Noting that "the tendency of some feminists to view the pill entirely as a male conspiracy seems unnecessarily crude,"[9] one author correctly identified women's desire to control their fertility and their important role in the acceptance of the pill in the early 1960s. Another writer rejected the earlier feminist stance that ignored women as actors, and argued that "to cast [women] in the passive role is to perpetuate the very kinds of assumptions about women that feminists have been trying to challenge."[10] This study attempts to situate events in the development and acceptance of the pill within the framework of the 1950s and 1960s in which they took place.

Three recurring themes run through this work. First, I consider the birth control pill to be a form of medical technology; as such, the issue of technological choice in the development, marketing, and use of the oral contraceptive is addressed.[11] The motivations of scientists and birth control advocates to create a new technological solution to the social problems of family planning and population control become clear when viewed through the lens of enthusiasm about science so pervasive in the 1950s. Similarly, women's rapid acceptance of the pill in the early 1960s must be considered in the context of contemporary attitudes toward technology and medicine. While skepticism about the benefits of the applications of the physical sciences grew in the 1930s and 1940s (particularly in the wake of the development of such war-related technologies as chemical weapons in World War I and atomic bombs in World War II), the biomedical sciences enjoyed a high level of public approval into the early 1960s as a result of the successes of wonder drugs such as antibiotics and the polio vaccine. As attitudes changed from optimism to skepticism in the late 1960s, faith in the pill also shifted to doubt and concern. Changing perceptions of the roles of technology and medicine in society affected the discourse on the safety and suitability of the pill for family planning and population control during the 1950s and 1960s.

The second thread that weaves through this history of the pill is the importance of the media in conveying information about science, health, and medicine to the public.[12] The birth control pill was at the boundary between the scientific and medical communities and the general public, and journalists mediated the flow of news from the former to the latter. Many people learned about the pill by reading articles in newspapers or magazines or by listening to news reports on radio or television.[13] These media sources also served as the public's primary means of information about the social, moral, economic, and medical effects of the oral contraceptives. Changes in the tone and content of print articles reflected shifting interests in the pill, from its novelty as the first new contraceptive in more than fifty years, to its influence on the moral fabric of society, to its health risks for individual users. By amplifying different issues concerning the pill as the decade progressed, the popular press helped to shape and re-shape public opinion of oral contraceptives during the 1960s.

The third theme traces the enduring power of the medical profession and the pharmaceutical industry in spite of persistent challenges to their authority.[14] From the early 1960s, sales in oral contraceptives added significantly to the drug industry's profits, and neither negative publicity nor lawsuits nor Senate hearings significantly diminished the commercial success of the pill. When the FDA proposed a package insert to warn patients about the possible adverse health effects of oral contraceptives in 1970, the pharmaceutical lobby exerted control over both the length and wording of the pamphlet. The medical profession also exercised its muscle to reduce the scope and content of the insert. Throughout this period, physicians jealously guarded their positions of authority in relationships with patients. Having arrogated birth control to themselves as a medical service requiring medical supervision, doctors were unwilling to yield control in this realm. In spite of concerns over health effects and the lack of adequate information on the relative risks and benefits of oral contraceptives, millions of women continued to visit their physicians each year to obtain pill prescriptions. Both the medical profession and the pharmaceutical industry successfully weathered the storm of concern about the safety of the pill in the late 1960s.

Organization of the Book

Ironically, neither the medical profession nor the pharmaceutical industry at first expressed interest in the scientific or commercial development of an oral contraceptive. Chapter 1 describes the context for the development of an oral contraceptive in the postwar years, when birth control still provoked controversy in American society and not everyone agreed on the need for a better contraceptive. Chapter 2 looks at the different kinds of information that promoted the acceptance of the pill by the medical profession and by women of reproductive age. It describes the changes fostered by doctors and patients in the delivery of family planning services and in the doctor-patient relationship. Chapter 3 focuses on perceptions of the pill at the specific moment in the mid-1960s when the popular press framed the discussion of oral contraception in social, moral, and political terms. Chapter 4 examines the evolution of the controversy over the adverse health effects of oral contraception and assesses the roles of physicians, patients, and the media in the increasingly public drama about its safety in the late 1960s. Chapter 5 uses the 1970 Senate hearings and the development of the FDA pill package insert to interpret the pill as a catalyst in the rise of two new movements: health feminism and informed consent. The final chapter takes a longer view of the pill in American society, discussing the consequences of the medical and popular commotion over oral contraceptives in the 1960s in light of the pill's record during the following three decades.

What makes the pill exceptional is that it was the first new method of birth control developed in the modern era. All other methods available in the 1950s had antecedents that dated back to ancient times. Intrauterine devices, vaginal suppositories and pessaries, douches, condoms, withdrawal prior to ejaculation, periodic abstinence based on the menstrual cycle, surgical sterilization, abortion, and infanticide had been used in different cultures for centuries. (Although abortion and infanticide have been and continue to be used as birth control, they are not considered here. This book focuses on birth control methods designed to prevent conception.) Women also ingested all sorts of concoctions to try to prevent conception or to abort a fetus, but these potions were based on magic, folklore, or just plain hope. The hormonal contraceptive developed in the 1950s was based on an understanding of the physiology and biochemistry of reproduction. Of course, it could not offer 100 percent protection; pregnancies could and did occur in women on the pill. However, the 98 or 99 percent effectiveness of the pill was considerably greater than that obtainable with any other contraceptive device or practice. The other methods, in spite of their lower rates of efficacy, represented age-old attempts to control fertility. The oral hormonal pill introduced highly reliable contraception and, for the first time, made voluntary pregnancy a real possibility for women.

The story of oral contraceptives, however, is about more than the development and distribution of a new method of birth control. It reveals much about the evolution of gender relations, particularly the professional relationship between women and their doctors. The debate over the pill's safety posed larger questions about the roles of physicians and patients in health care and helped to produce a new feminist ideology of the body, particularly with regard to reproductive health. This study of a medical technology designed to meet a social need affords an opportunity to examine attitudes toward sex and sexuality, women's health and medicine, and science and technology in late-twentieth-century American culture. Ultimately, the history of the pill compels us to reassess the medical applications of science in our lives.

Chapter One

Genesis of the Pill

American Women in the 1950s

After years of economic depression and war, Americans in the late 1940s and early 1950s leapt eagerly into domesticity and consumerism. They married, had children, and set up households as soon as they could. During the 1950s, the marriage rate reached an all-time high and the average age at which people married dropped to an all-time low. By 1959, almost half of all brides walked down the aisle before their nineteenth birthday. College women joined the trend, too: many married during their student years, while others dropped out to become full-time wives. An advertisement for Gimbel's department store summed up the 1950s attitude toward higher education for women: "What's college? That's where girls who are above cooking and sewing go to meet a man so they can spend their lives cooking and sewing."[1] These newlywed women soon found themselves cooking and sewing for three; half of them got pregnant within seven months of their weddings. The new parents of the postwar era did not stop at one child or even two, so that between 1940 and 1960 the birth rate doubled for third children and quadrupled for fourth children.

These middle-class, mostly white families bought homes in the new suburban developments, from which the men departed each morning to their jobs and within which the women busied themselves with housework and childcare. The new homes needed to be filled with furniture, appliances, and the latest conveniences for modern living. The family needed to have two cars, one for dad to drive to work and one for mom to chauffeur the children and to do errands. These material acquisitions required money, not all of which could always be provided by the husband. To be able to afford all of the domestic luxuries available, many women took jobs outside the home. Between 1940 and 1960, the number of working women doubled and the number of working mothers quadrupled; by 1960, 30 percent of married women and 39 percent of women with children held jobs. It was ironic that a large proportion of the women entering the postwar workforce were the same middle-class wives and mothers who supposedly had found new satisfaction in homemaking.[2]

Indeed, women who worked went against the tide of professional advice and public opinion that dictated women belonged at home. Magazines popularized antifeminist doctrines in articles that glorified the life of a housewife. *Seventeen* magazine preached the gospel of domesticity to teenagers: "there is no office, lab, or stage that offers so many creative avenues or executive opportunities as that everyday place, the home . . . What profession offers the daily joy of turning out a delicious dinner, of converting a few yards of fabric, a pot of paint, and imagination into a new room? Of seeing a tired and unsure man at the end of a working day become a rested lord of his manor?"[3] This article neglected to excite its readers about impending motherhood and the thrill and challenge of changing diapers, preparing bottles, and spending all day every day alone in that well-decorated house with infants and toddlers for company.

No wonder then that some educated women chose to work, claiming the socially acceptable reason of making an economic contribution to the family's well-being. Of course, not all of the women in the labor force took jobs to earn money for a new dishwasher. Many women worked to pay for the necessities of life. However, among the white middle class for whom luxuries were within reach, the wife's job enabled the family to attain a higher standard of living. Of these women who elected to work, many enjoyed their jobs; employment offered women not only paychecks but also the opportunity for social interaction and personal fulfillment. Work may have offered a solution to what Betty Friedan called "the problem that has no name," the boredom and dissatisfaction experienced by young wives and mothers across the country.[4]

Women did not find it easy to juggle the responsibilities of work, home, and family. Women who worked outside the home could not afford the time and energy to have more and more children. Nor could those who stayed home all day afford the financial, emotional, or physical strain of having babies every year or two. A number of factors conspired in the postwar years to encourage women to have several children. The relative economic prosperity of the late 1940s and 1950s promoted larger families than those of the previous Depression-era generation. For many middle-class women, personal success was to be found at home, in contrast to the achievement sought by their husbands in the workplace. If motherhood was a full-time career, then women could extend the period of fulfillment by having more children. Nevertheless, in spite of the pronatalist climate fostered by psychologists, physicians, and other experts and popularized by women's magazines, most women did not plan or desire to have six or eight or ten children.

At first glance, it may seem contradictory that women who had several children also wanted effective contraception. However, many of those women who fit their lives into the culturally approved mold of the American housewife later became discontented with the role of suburban wife and mother. Furthermore, women who married early and completed their families by the age of thirty still needed birth control for the remaining fifteen or more years of their reproductive lives. Whether they worked for luxuries, or out of necessity, or not at all, married women in the 1950s, like previous generations, needed birth control both to limit their total number of children and to space the children they chose to have.

Birth Control in the 1950s

What contraceptive options did women have in the 1950s? The most effective method of birth control was a diaphragm used in combination with spermicidal jelly. The next most effective method, the condom, could be purchased at the drugstore. Other commonly used methods—withdrawal, douching, and rhythm—were relatively fallible. Although in theory the diaphragm and the condom effectively prevented pregnancy, in practice they were less reliable. In order to obtain a diaphragm, a woman had to ask her physician to fit and prescribe one for her, which meant that in essence she had to ask him for permission to plan her family.[5] Condoms required the cooperation of the male partner, and both methods entailed touching the genitalia, which many people found troublesome. Many couples used one or more methods of contraception with varying degrees of success.

By the late 1950s, the large majority of married people, with the exception of Roman Catholics, accepted and used birth control. Most Americans were not troubled by the moral implications of contraception and the separation of sexual intercourse from procreation.[6] The problem was that the current methods of birth control offered neither high efficacy nor convenience. Nonetheless, women made do with what was available. Some couples, particularly the more affluent, would not have despaired if birth control failed a couple of times and they ended up with five children rather than three. Or, if a woman did despair, she kept it to herself, and instead paid lip service to the cult of domesticity and motherhood. In the restrained culture of the 1950s, ordinary women did not speak out for the development of a better contraceptive.

In those years, marriage was a prerequisite for sex and pregnancy. Society frowned upon women who engaged in premarital sex; nevertheless, some women chose to ignore social mores. These women faced the double threat of unwanted pregnancy and the disapproval of family and friends. Even so, very few of them used contraception. Not only was it difficult for single women to obtain birth control, its use implied premeditation, which was unthinkable in the context of the times. One woman interviewed for an oral history of the 1950s recalled: "It never entered my mind to use birth control. It never really occurred to me that I could get pregnant. I knew birth control existed, but I didn't know anything about it. To go out and actually get it would mean that I planned to do these things, to have sex. Since I knew it was wrong, I kept thinking I wasn't doing it, or I wasn't going to do it again. Each time was the last time. Birth control would have been cold-blooded."[7] This woman used denial to deal with the contemporary taboo against premarital sex; if she did not plan to have sex, then she certainly did not plan to get pregnant, so she did not need to consider, much less use, contraception.

State laws against contraception and abortion lent support to society's simultaneous encouragement of childbearing for married women and disapproval of sexual activity for unmarried women. As late as 1960, thirty states had statutes prohibiting or restricting the sale and advertisement of contraceptives.[8] In Connecticut, the use of birth control drugs or devices was illegal and punishable by a fine of $50 and/or a prison sentence of up to one year. In 1965, the U.S. Supreme Court ruled Connecticut's law unconstitutional because it violated a married couple's right to privacy.[9] The right of privacy in matters of birth control was not fully extended to the unmarried population until 1972, when the Supreme Court declared unconstitutional a Massachusetts law prohibiting the sale of contraceptives to unmarried people.[10] By making information about and access to birth control difficult to obtain, the laws on the books in the 1950s effectively prevented many women from controlling their fertility.

Most of the state laws on birth control allowed contraceptives to be sold only by physicians or pharmacists acting on a physician's prescription. However, a 1957 study of doctors' attitudes and practices regarding contraception found that women were reluctant to ask their doctors for contraceptive information and advice, and that doctors, too, hesitated to offer family planning services unless specifically asked by their patients.[11] Seventy percent of the non-Catholic physicians (and 83% of the Catholics) thought that family planning should be a supplementary medical service at the request of the patient, as opposed to a regular procedure offered by a physician. Half of the doctors in the study said that they never introduced the subject of birth control during their medical examinations of premarital patients; the same proportion never, or hardly ever, brought up the subject with their postpartum patients. However, the reluctance of physicians to discuss family planning with their patients did not mean that women were kept ignorant of birth control; instead, they consulted other sources for information and advice. Physicians recognized the importance of these alternative sources: 58 percent of those interviewed figured that most women in their communities learned about birth control from friends (only 27% named other physicians as the primary source of information).

Although the American Medical Association had officially sanctioned birth control as a part of medical practice twenty years earlier, this study demonstrated the lingering ambivalence of physicians toward their role in family planning. In the 1950s, the only unique service that a private physician could offer in the realm of contraception was to fit diaphragms; condoms, douches, and spermicidal jellies could be purchased over the counter at the drugstore or at family planning clinics, and advice could be sought over the back fence. A small proportion of doctors worked in family planning clinics, but the majority of private physicians gained little by offering birth control services to their patients: the work was not financially rewarding, medically challenging, or professionally acknowledged.

So physicians did not encourage women to use contraception, and state laws effectively discouraged women from seeking birth control information and devices. Society reinforced these policies in magazine articles and advertisements geared toward the homemaker. Who, then, identified the need for an improved contraceptive, one that would be safe, convenient, and reliable, and pushed for its development? One of the few voices that rang out during the baby boom years of the late 1940s and 1950s belonged to Margaret Sanger, for forty years America's most outspoken advocate of birth control.

Margaret Sanger, Birth Control Champion

Margaret Sanger opened the first birth control clinic—illegally—in Brooklyn in 1916.[12] In the ensuing years, she worked tirelessly to advocate the legalization of contraception and to promote the development of clinics where women could obtain safe and effective contraceptives. In 1936, the U.S. Circuit Court of Appeals modified the fifty-year-old Comstock Act by removing contraceptives from the list of obscene materials, thus allowing birth control information and devices to be mailed to and distributed by physicians. Although Sanger focused her attention in the postwar years on the founding of the International Planned Parenthood Federation, she retained her interest in research efforts to improve existing contraceptive methods and to develop new ones. She realized that in order to be used effectively and consistently, a contraceptive had to be safe, simple, inexpensive, and controlled by women.

In her earlier days as a nurse working in the ghettos of New York City, Sanger witnessed the poverty and despair of mothers with too many children. She had always championed the right of women to choose whether and when to have children, and she argued that the only way to achieve voluntary motherhood without abstinence was to have access to birth control information and methods. Since women became pregnant, she reasoned, they should be in charge of contraception. In the late 1940s, Sanger became increasingly dissatisfied with the diaphragm and spermicidal jelly method of contraception, the method she had advocated for the past thirty years, because it appealed to a limited audience of women worldwide. Millions of women lacked the sanitary facilities necessary for hygienic use of the diaphragm; others disliked the intimate physical contact it required. Sanger envisioned a "birth control pill" to simplify the practice of contraception.[13] In the 1950s, she lobbied for an oral contraceptive because a pill that could be taken at a time and place independent of the sex act would place the control of contraception solely in the hands of women, where it belonged. At the same time, she held to her conviction that birth control should remain fully within the purview of the medical profession, so that pills would be available only with a doctor's prescription.

Sanger promoted the medical profession as the "proper authority" for dispensing contraceptive information and devices. In the 1920s and 1930s, she figured that Congress was more likely to legalize contraceptives as prescription-only devices than to pass legislation that made birth control an unrestricted commodity. She did not champion the medical profession for political reasons only. Her trust in organized medicine paralleled her faith in science and technology; she believed in the potential of scientists to develop a solution to the problem of a better contraceptive. Sanger was confident that once scientists turned to the task of finding improved methods of contraception, they would succeed; she worked to focus their attention toward this goal. In speeches, she exhorted doctors and scientists to take up the challenge: "In my opinion the proper authorities to give advice on Birth Control are the DOCTORS and NURSES. No other class of men or women are so AWARE of the NEED of this knowledge among working people as they. YET THEY HAVE REMAINED SILENT! The time has come when they MUST TAKE THEIR STAND in the progressive movement. For though the subject is largely social and economic yet it is in the main physical and medical, and the object of those advancing the cause is to open the doors of the medical profession, who in turn will force open the doors of the laboratories where our chemists will give the women of the twentieth century reliable and scientific means of contraception hitherto unknown."[14] For Sanger, as for many of her contemporaries, science and technology held the key to a better future. She firmly believed that the social problem of birth control could be solved by the application of science, so long as time and funds were allocated to the effort. Sanger lived in the age of scientific miracles; not unreasonably, she expected that her cause would also benefit from twentieth-century scientific progress.

Sanger's dedication to the birth control movement stemmed not only from her concern for the well-being of individual mothers and their families but also from a larger concern about the societal impact of too many unwanted children. On her travels to the Far East in the 1920s, she witnessed what she interpreted as masses living in poverty in Japan, China, and Korea, and concluded that a nation's ability to control the fertility of its people advanced its economic development. While Sanger found little support during the late 1940s for her advocacy of American women's right to plan and to limit family size, her concern about rapid population growth was shared by other Westerners, who expressed alarm about the implications of an overpopulated world.

Population Control

The roots of population control can be traced back to the Reverend Thomas Malthus, who wrote *An Essay on the Principle of Population* in England at the end of the eighteenth century.[15] Malthus pointed out a discrepancy between the rate of growth of a population and the rate of growth of its food supply. According to Malthus, since the amount of food increased arithmetically, that is at a fixed rate (a statement since proven incorrect), the potentially geometric rate of population growth had to be controlled by either positive or preventive checks. The former consisted of premature deaths from causes such as disease, war, and famine; the latter consisted of the forestallment of births by delayed marriage and abstinence. Malthus did not include birth control as a preventive check on population growth; he considered the notion of contraception to be abhorrent. His gloomy forecast of the ebb and flow of population growth was not merely a theoretical treatise; Malthus used his thesis to argue against charity in the form of England's "poor-laws."

In the early twentieth century, the eugenics movement in America played on similar fears of a teeming underclass.[16] Eugenicists drew attention to differential fertility rates of different social classes and ethnic groups within the United States and expressed alarm that the upper-class, native, white population had a lower birth rate than the lower-class, immigrant population. Many who embraced the ideology of eugenics in the 1910s and early 1920s sought to restrict immigration into the United States. In 1924, their efforts resulted in the passage of the Johnson Act, which set strict limits on the number of immigrants from countries other than those of northern Europe. Eugenicists also proposed to counteract the alleged "race-suicide" by encouraging those they deemed "fit" (middle- and upper-class whites) to have lots of children (positive eugenics) and those they deemed "unfit" (immigrants, the poor, the handicapped) to control their fertility (negative eugenics). This latter goal could be achieved more directly and more permanently by sterilization and the interdiction of "dysgenic" marriages than by the use of contraception. Most eugenicists disapproved of birth control because they believed that educated women used such methods disproportionately, further contributing to differential fertility rates among social classes. Eugenicists were not particularly interested in maternal and child welfare, so they did not consider contraception a necessary or valuable tool. Nor did they pay much attention to the prospect of overpopulation; only after World War II did that issue come to occupy center stage.

After 1945, a new generation of demographers shifted their focus beyond America's borders to the international scene, where they identified a trend of rapid population growth in underdeveloped nations. As a result of simple technological, relatively inexpensive public health measures introduced by various aid agencies, these countries had experienced decreases in death rates without corresponding decreases in birth rates. The new advocates of population control assumed that overpopulation would hinder economic development, which in turn could lead to political instability. In the context of the Cold War, American strategists considered it vital to foster economic progress in the capitalist tradition within developing countries in order to prevent their defection to the Communist bloc.[17]

In spite of the warnings from demographers, neither the U.S. government, nor the United Nations, nor the major foundations were willing to include family planning in their programs. Western nations and philanthropic organizations did not want to be accused of political or cultural insensitivity in their efforts to boost the economies of other countries. Birth control was not only a delicate issue because it pertained to sex, but was also extremely offensive to Catholics. In the 1950s, the Rockefeller Foundation rejected fertility and family planning research in favor of the less controversial route of agricultural research—eventually, the "Green Revolution"—to accommodate the growth in world population rather than to control it.[18]

The Population Council

One of the Rockefellers, however, refused to sit by while the population continued to grow unchecked. In 1952, John D. Rockefeller III arranged a conference of thirty-one scientists and scholars to discuss "population problems." The group met in June in Williamsburg, Virginia, under the auspices of the National Academy of Sciences and concurred that rapid population growth presented problems not only in food supply and distribution but also "in relation to all the things needed for a full life." Much of the discussion focused on Third World countries. Turning the usual course of events upside down, several participants concurred that "in view of the demographic situation in India . . . it would be important to achieve a reduction of the rural birth rate prior to the transition to an industrial and urbanized society."[19] In other words, instead of waiting for social and economic improvements to lead people to control their fertility voluntarily (as had been the case in western Europe and the United States), the experts wanted to introduce birth control before modernization and industrialization took place. They realized that the methods by which this plan could be implemented were by no means obvious, and they agreed that much more research was needed on the physiological, psychological, and cultural determinants of human fertility.

The conference participants recognized that Americans should not play a dominant role in fertility control in other countries, but rather should offer assistance in research, education, training, and publicity. Some of the conferees acknowledged that "the discussions at the conference had represented a particular point of view—which might be called the Western Protestant point of view. In the world as a whole, this was a minority point of view."[20] Nonetheless, most believed that they were right about the dismal economic forecast produced by overpopulation, and their intentions to foster economic development were genuine, however much impaired by the narrow focus on fertility control, to the exclusion of other relevant social factors.

What is interesting about the recommendations of this group and about the mission of the Population Council to which it led, is the expression of unequivocal faith in science and technology to deal with the crisis in population. They felt confident that scientific research would help to solve the perceived problem of a population explosion. Whether through new contraceptive methods or new agricultural methods, research would lead to technological solutions. In this way, the advocates of population control shared Margaret Sanger's optimism and enthusiasm about the potential of science and technology to help solve social problems.

In November 1952, John D. Rockefeller III contributed $100,000 to finance the Population Council, a nonprofit, private organization with an international focus on population and related concerns. An elite group of men in the sciences and social sciences constituted the initial board of trustees: Detlev Bronk, president of the Rockefeller Institute and president of the National Academy of Sciences; Karl T. Compton, chairman of the corporation of the Massachusetts Institute of Technology; Frank Notestein, director of the Office of Population Research at Princeton University; Frederick Osborn, secretary of the American Eugenics Society; Thomas Parran, dean of the Graduate School of Public Health at the University of Pittsburgh; John D. Rockefeller III; and Lewis Strauss, later chairman of the U.S. Atomic Energy Commission. The reputations of these men as national leaders in the sciences not only accorded status to the new organization, but also prevented condemnation by birth control opponents.[21]

From the outset, the trustees of the Population Council did not see the objectives of the organization as including birth control education and dissemination; those activities remained within the domain of the national birth control organization, the Planned Parenthood Federation of America. According to John D. Rockefeller III, "We all agree that the direct promotion of the use of contraceptives is not our responsibility [though] we are engaged in the scientific study of contraceptive methods."[22] During the 1950s, the Population Council operated in a behind-the-scenes fashion, facilitating research in the two main areas of demography and reproductive physiology by soliciting contributions from foundations and philanthropists and awarding grants and fellowships to individual researchers and laboratories. In this way the Population Council served as a clearinghouse for both information and funding in population research.

The biomedical division of the Population Council sponsored basic research in human reproductive physiology on such topics as sex hormones, spermatogenesis, the physiology of fallopian tubes, the chemistry and physiology of cervical mucus, and immunology. The division financed projects they believed would contribute to the development of new contraceptive methods. Once again, science was enlisted in service to humanity; the objective of this arm of the Population Council was to direct scientific research toward the development of a technology that could be used to deal with the social and economic problem of fertility control.

Despite this mission and its support of a variety of basic science projects, the Population Council did not become involved in the investigations that eventually resulted in the oral contraceptive pill. That work received initial funding from the Planned Parenthood Federation. Formed in 1942 out of the old Birth Control Federation of America, Planned Parenthood stitched together the birth control movement and the population control movement into a virtually seamless fabric in the 1950s and 1960s, so that most Americans came to use the terms "birth control," "family planning," and "population control" interchangeably.[23] The national committee served as the umbrella organization for more than a hundred birth control clinics around the country; it also engaged in active fundraising and grantmaking to research groups, in much the same way the Population Council did.

Planned Parenthood's leaders often interacted with those of the Population Council, and they shared the same concerns about overpopulation in the 1950s and 1960s. Birth control and population control proved to be compatible goals, so the two organizations worked in a complementary fashion. Planned Parenthood's main service was the dissemination of contraceptives; its objective was to improve family life by enabling couples to plan when to have children and how many to have. Planned Parenthood embraced the idea of an ideal family size based on the upper middle-class values of its leadership.[24] While the clinics offered patients the range of available methods, the national committee actively supported research into possible new methods that would reach a broader segment of the population, particularly women of low socioeconomic status. Planned Parenthood, like its founder, Margaret Sanger, believed that the answer to many of the problems of families and society lay in the development and distribution of a better contraceptive.

The Pill and Technological Choice

At this point, it is instructive to try to answer the question, Why the pill? That is, what factors influenced the choice of this particular contraceptive technology as opposed to other possible methods? The climate of the 1950s favored scientific and technological solutions to social problems. In the early part of the decade, for example, most Americans still approved of the use of the atomic bomb to end World War II; they did not yet understand the full implications of nuclear weapons and technology. In the field of medicine, scientists working in the laboratory produced antibiotics, the cure for bacterial infections, and Americans eagerly anticipated the next wonder drug. Given the popular optimism and enthusiasm for science and its products, it made sense that birth control advocates would look to science for a new contraceptive technology instead of focusing their efforts on public education or the improvement of existing barrier contraceptive methods. The bet that the public would be receptive to a new method of birth control was a good one.

Intellectual, economic, and sociocultural factors conspired to preclude the development of a male contraceptive. In the first place, scientists were daunted by the task of inactivating the millions of sperm produced by the human male each day; instead, they focused their efforts on preventing the female's monthly release of a single ovum. Second, Margaret Sanger and her colleague Katherine McCormick insisted on a method that could be controlled by the woman; they both felt that birth control should be the right and responsibility of women. McCormick commanded attention because she provided most of the financial backing for the development of an oral contraceptive. She decided to use her inheritance to support a hormonal pill and not any other potential contraceptive.[25] Third, men were averse to the idea of physiological control of their reproductive systems. One scientist commented dryly on the dearth of male volunteers for contraceptive research: "Male volunteers for fertility control studies may be numbered in the low hundreds whereas women have volunteered for similar studies by the thousands . . . he [the human male] has psychological aversions to experimenting with sexual functions . . . perhaps experimental studies of fertility control in men should be preceded by a thorough investigation of male attitudes."[26] Finally, social convention echoed the sentiments of Sanger and McCormick: contraception, like pregnancy and childrearing, was considered to be a female responsibility. Men could dictate the circumstances of these "feminine" activities (e.g., the appropriation of childbirth by the medical profession), but the responsibility remained solidly within women's social sphere.

Science could have offered several possible avenues for the development of a new contraceptive. Research focused on steroid hormones instead of other possible methods, such as an antisperm vaccine, because scientific knowledge in the areas of steroid chemistry and female reproductive biology had been advanced in recent years by several important developments.[27] In the early 1950s, research in reproductive physiology, endocrinology, and steroid chemistry came together to lay the foundation for applied research toward a hormonal contraceptive. Work on a vaccine never took off because the basic science of immunology was not as well understood.

The availability of a scientist experienced with both steroids and reproductive physiology also directed the path of research toward a new contraceptive. Margaret Sanger's call for a scientist to investigate the possibility of a birth control pill was answered by Gregory Goodwin Pincus, a biologist at the Worcester Foundation for Experimental Biology in Shrewsbury, Massachusetts. After their initial meeting in late 1950 or early 1951 through a mutual friend, Dr. Abraham Stone, director of the Margaret Sanger Research Bureau in New York, a year later Sanger and Pincus began to talk seriously about research into hormonal contraception. Pincus had professional experience in both reproductive science and hormone research. During the early part of his career, he studied the process of fertilization in mammals and other aspects of mammalian sexual physiology. In the 1940s, at the Worcester Foundation, he engaged in research sponsored by G. D. Searle and Company to devise a method for large-scale production of cortisone, a steroid hormone used to treat arthritis. Pincus's background and position at the Worcester Foundation suited him well for work on an oral contraceptive.[28]

Thus, a combination of scientific, economic, social, and cultural forces acted in concert to drive contraceptive research in the direction of a hormonal pill for women. The state of scientific knowledge in biology and chemistry, the passion and conviction of Katherine McCormick and Margaret Sanger, the cultural reluctance in America to deal openly with sex and sexuality, and the public's faith in the ability of science and technology to solve social problems dictated the path of contraception research in the 1950s. Spurred by one determined feminist and sustained by another, biologist Gregory Pincus used his experience in the advancing field of reproductive endocrinology to create a scientifically feasible and culturally appealing solution to the problem of family planning. The observation that "social institutions shaped women's technological options" in birth control certainly holds true for the development of the pill in the 1950s.[29]

The Scientific Background

The study of reproduction did not emerge as a scientific discipline until after 1910. By that time, biologists had recast nineteenth century inquiries about heredity, development, and evolution into the twentieth century subjects of genetics, embryology, and reproductive science.[30] Economic and social factors (i.e., the availability of funding, the possibility of applications in agriculture and medicine, and the rise of the eugenics and neo-Malthusian movements) were important in the establishment of the reproductive sciences in the period 1910–40.[31] Concurrent progress in the new field of hormone research also spurred studies of reproductive physiology, as scientists discovered the importance of chemical messengers in the mammalian reproductive cycle. Although researchers in the 1930s and 1940s did not explicitly frame their studies in terms of birth control, their work provided the groundwork for Pincus and his colleagues at the Worcester Foundation in attacking the specific problem of finding a suitable oral contraceptive.[32]

During this "heroic age" of reproductive endocrinology, biologists studied both the roles of the steroid hormones and the relationships between the structures of reproductive organs and their functions.[33] They solved the riddle of the female reproductive cycle, demonstrating the effects of changes in hormonal levels on the physiology of reproductive structures.

Briefly, at the start of the reproductive cycle, the pituitary gland at the base of the brain secretes follicle-stimulating hormone (FSH), which stimulates the follicles of the ovary to mature. These ovarian follicles secrete the hormone estrogen, which then stimulates the lining of the uterus to thicken and to become enriched with blood vessels. The pituitary responds to the increasing level of estrogen in the blood by decreasing the secretion of FSH and then secreting luteinizing hormone (LH), which induces a follicle to rupture and release an egg (ovulation). The ruptured follicle, now called the corpus luteum, secretes the hormone progesterone, which maintains the thickened uterine lining within which a fertilized egg could implant. So long as the corpus luteum is secreting progesterone, the pituitary will not secrete FSH, so other eggs are prevented from maturing. After its release from the ovary, the egg travels through the Fallopian tubes on its way to the uterus. If the egg is fertilized by a sperm, it will implant in the wall of the uterus. The fertilized egg then secretes a hormone called chorionic gonadotropin, which maintains the corpus luteum and its secretion of progesterone until the placenta forms and takes over progesterone production. If no egg implants, the corpus luteum breaks down. As the amount of progesterone in the blood decreases, the uterine lining breaks down (menstruation) and the pituitary increases the secretion of FSH to begin another cycle.

The pill works primarily by inhibiting ovulation. Synthetic estrogen and progesterone elevate the hormone levels in the blood, preventing the pituitary gland from releasing FSH, so no egg is stimulated to develop within the ovary. The synthetic progesterone component increases the thickness of cervical mucus, incapacitates sperm, slows the movement of the egg, and prevents complete development of the uterine lining. All of these effects provide important contraceptive backups in case hormone levels are not high enough to inhibit ovulation.

By 1940, scientists had succeeded in isolating both estrogen and progesterone and elucidating the chemical structure of each molecule. Physicians began to experiment with both hormones as therapeutic agents to treat painful menstruation and other gynecological disorders. However, it was both difficult and time-consuming to obtain the pure hormones from natural sources; for example, 1 ton of animal organs yielded only 1 gram of progesterone. This laborious purification process translated into a very expensive product: a single gram of progesterone cost $80 and could be afforded only by the wealthiest patients.[34] Then in 1943 the chemist Russell Marker devised a way to derive progesterone from diosgenin, a steroid compound found in the roots of the wild Mexican yam plant; this innovation helped to lower the cost of progesterone dramatically. Unfortunately, progesterone derived from either plant or animal sources was orally effective only when ingested in very large doses and thus remained clinically unfeasible.

In the 1940s, most research in the reproductive sciences took place in university laboratories. In the early decades of this century there were active centers of reproductive sciences in biology departments at schools across the country in which scientists investigated basic questions about the physiology of reproduction.[35] Although they may have had to defend their research in terms of potential practical applications in order to procure funding, reproductive biologists designed their experiments to study the fundamental processes of mammalian reproduction. By the 1930s, the scientific problems suggested by social concerns had gained legitimacy in their own right. Investigators and their funders began to base evaluations of research projects on human reproduction or scientific merit alone rather than the potential to solve social problems.[36]

By contrast, chemists, particularly those employed by the private sector, chose research topics with obvious practical and commercial applications. A significant proportion of research in chemistry was conducted in industrial laboratories in the 1940s. In that decade, chemists turned their attention to the synthesis of steroid hormones. Pharmaceutical companies, driven by the profit motive, supplied physicians with synthetic hormones for use in treating patients. Frank Colton, a chemist hired by G. D. Searle and Company in 1951, described the mission of the research division: "The overall objective of our steroid program was to discover, develop, and introduce into clinical use better steroid drugs than those available at the time and/or steroids that would be useful in conditions for which no compounds were previously available."[37]

In 1949, a race began among several drug companies to find an economical method of producing cortisone, which had been demonstrated to have almost magical therapeutic results in arthritic patients. One scientist recalled the impact of the discovery of this application of cortisone: "The therapeutic effectiveness of cortisone in rheumatoid arthritis hit the pharmaceutical industry and indeed the whole lay public like a bombshell. Newspaper articles featured crippled arthritics who were able to dance within a few days or weeks after cortisone therapy. Cortisone was hailed as a new miracle drug, which the pharmaceutical companies saw as a potential bonanza (to be used by millions of patients) and the public saw as a reaffirmation of its faith in the magic of modern medicine."[38] Upjohn ultimately beat out the competition by employing microbiological fermentation to manufacture cortisone. Cortisone was only one of several steroid hormones with therapeutic applications. Other drug companies quickly turned to the potentially lucrative group of sex hormones.

Researchers at two companies, G. D. Searle and Syntex, S.A., independently attempted to make a "better" progesterone, stronger than the natural hormone and able to be administered orally. Neither group set out to create a contraceptive drug; instead, they hoped to capitalize on the therapeutic value of progesterone in various gynecological disorders.[39] In 1951, Carl Djerassi and his collaborators at Syntex succeeded in synthesizing a compound called 19-nor-17α-ethynyl-testosterone, better known as norethindrone, an orally active progestational hormone. The following year, Searle's Frank Colton developed an alternative procedure to synthesize norethynodrel, another orally active progestational hormone that differed in structure from norethindrone only in the location of one of the carbon double bonds. On November 22, 1951 Djerassi applied for a patent for his compound and the procedure by which it was produced; the patent was granted on May 1, 1956. Colton applied for his patent on August 31, 1953; that patent was awarded on November 29, 1955. Both synthetic hormones received FDA approval for the treatment of certain gynecological disorders in 1957, but the first oral contraceptive contained Searle's progestin, norethynodrel. That circumstance depended less on the merit of the two different compounds than on the willingness of different pharmaceutical companies to involve themselves in selling birth control pills.

The Pharmaceutical Industry

The drug industry shied away from contraceptive research and development in the 1950s because it did not want to challenge the anti–birth control laws still active in several states and because it feared boycotts of their other products by indignant Catholics.[40] Recall that the majority of states still had laws regulating the sale and advertisement of contraceptives. No business wanted to market a new product under such limiting conditions. Nor did it want to incur the wrath of the Roman Catholic Church, which could influence the purchasing decisions of its laity (Catholics made up 25 percent of the American population in 1960). Parke-Davis, Charles Pfizer, Upjohn, and Ortho (a subsidiary of Johnson & Johnson) all passed on the opportunity to produce the world's first birth control pill. Syntex did not have a retail marketing division in the United States, and thus relied on contracts with other pharmaceutical companies to handle its products. Although Syntex had plenty of progestin to sell, it could not enter the market until it found a willing American partner. The story of how Searle became the manufacturer of the first oral contraceptive can be traced back to the company's long-standing association with Gregory Pincus.

Searle had retained Pincus as a consultant since the early 1940s. The company hoped that Pincus and his steroid research team at the Worcester Foundation would devise a method for synthesizing cortisone on an industrial scale. The scientists succeeded in producing mass quantities of cortisone (by perfusing animal adrenal glands with a precursor steroid compound and letting the organs do the work), but only at a very high cost. When Upjohn revealed its inexpensive microbiological method, Searle had to capitulate.

Although Searle continued to sponsor research at the Worcester Foundation, the company expressed its disappointment with Pincus and refused to underwrite his early investigation into the feasibility of a hormonal contraceptive. However, later, when Pincus wanted to substitute synthetic progestin for natural progesterone in his experiments, Searle supplied him with several possible compounds. The company routinely sent the scientist their latest chemical products so that he could determine whether they had any biomedical, and thus commercial, value.

By contributing drugs, not significant funds, to Pincus's research, Searle kept its options open with regard to a potential birth control pill without making a significant financial investment. Only when Pincus had proven his success in developing a contraceptive pill did Searle make a serious fiscal commitment to the project; in 1957, the company purchased a Mexican supplier of plant steroids (the starting material for norethynodrel). The following year, Searle's director of clinical research wrote to Pincus: "It is no news, I am sure, that the powers that be are breathing down our neck in the hopes of speeding up our application to the Food and Drug Administration on the contraceptive utility of Enovid."[41] By 1958, Searle was eager to scoop its competitors and did not want to squander its advantage over Syntex, whose difficulties in negotiating a deal with an American manufacturer delayed the development of norethindrone into a marketable pill.[42]

Funding the Pill Project

If Searle did not make a major financial contribution to Pincus's pill research, then who did? Searle awarded grant monies to the Worcester Foundation every year, as did several other pharmaceutical companies, but the combined total from the drug industry averaged only $111,000 per year during the decade 1951–61, and that sum was distributed among numerous research projects at the foundation.[43] The federal government did not get involved; despite the increasing support of scientific research by the National Science Foundation and the National Institutes of Health, neither agency funded any investigations in the reproductive sciences in the 1950s.[44] The Planned Parenthood Federation gave Pincus a grant of

$3,100 in 1951 to study hormonal contraception, which was renewed in 1952 ($3,400) and again in 1953 ($3,600), but these sums barely paid the salary of one researcher.[45] Planned Parenthood did not have the financial resources to support Pincus's project, which would cost hundreds of thousands of dollars, but the federation expressed a keen interest in the potential application of his research.

The Population Council did deal in significantly larger amounts of money, but when Pincus began his contraceptive research, the council had not yet been formed. He wrote to the new organization in 1953 about the possibility of obtaining funding, and in 1954 the Worcester Foundation received a grant of $10,000 for the study of the physiological and biochemical changes of sperm in the female reproductive tract, not for the oral contraceptive project.[46] By that time, the Population Council knew that Pincus had found a more than adequate source of funds for his research. For the duration of the pill project, the financial contributions of the pharmaceutical industry and the birth control organizations would be eclipsed by the donations of a single individual, Katherine Dexter McCormick.

Born in 1875 to a prominent Chicago family, Katharine Dexter solidified her position of wealth in 1904 by marrying Stanley McCormick, scion of the head of the International Harvester Company.[47] Mrs. McCormick played an active role in the woman's suffrage movement, donating both her time and money to the cause. She met Margaret Sanger in 1917 and within a few years began to contribute to the birth control movement. However, contraceptive research was just one of McCormick's interests in science; she also invested heavily in research on schizophrenia, because her husband had fallen victim to the disease just two years after their marriage. For twenty years (1927–47) she funded the Neuroendocrine Research Foundation at Harvard in a vain attempt to find a cure for schizophrenia. Upon her husband's death in 1947, Mrs. McCormick ceased her support of this research.

As the story goes, McCormick renewed her acquaintance with Sanger in 1950. In 1952, the two women met to discuss how McCormick could best contribute to research into "a fool-proof contraceptive."[48] By this time, Sanger had discussed the possibility of a birth control pill with Gregory Pincus and knew that he had begun his initial investigation into the problem. Sanger and McCormick visited Pincus at the Worcester Foundation on June 8, 1953, at which point McCormick pledged to support the pill project through its successful completion. Her initial gift of $10,000 grew to an annual contribution of $150,000–$180,000 until her death in 1967. At first McCormick channeled her money through the Planned Parenthood Foundation, but she became increasingly frustrated with the organization's management and later contributed directly to the Worcester Foundation.

McCormick's personality and personal situation endowed her with a strong sense of responsibility for the projects she supported. It is equally as important to consider McCormick in the context of her times. Graduating from the Massachusetts Institute of Technology in 1904 as one of only a few women to receive degrees from that university, she came of age amidst the new emphasis on experts, professionals, and science during the Progressive Era. When she turned her attention to the problem of contraception in the 1950s, she faced two sets of "experts": the birth control administrators at Planned Parenthood and the scientists at the Worcester Foundation. McCormick, like Sanger, believed that science and technology held the answer to the social problem of birth control and thus she threw the full weight of her support behind Pincus and his research team. Indeed, Planned Parenthood executives also appreciated the importance of the scientist, as evidenced by this letter to Pincus from the national director: "In what specific ways can you anticipate . . . that the results of your research may be put to work? . . . a constant problem in most research is selling the idea of research itself, especially to those who may help support it. No one can do this as authentically as the scientist."[49] In 1965 Pincus dedicated his book, *The Control of Fertility,* to Katherine McCormick "because of her steadfast faith in scientific inquiry."[50]

In addition to her faith in science and the scientist as expert, McCormick remained committed to her belief in women's rights. McCormick had spent her young adulthood on the front lines of the women's suffrage movement. She believed that the right to reproductive control was as important as the right to vote. For McCormick (again, like Sanger), birth control belonged in the hands of women. Unlike later feminists, these two saw female-controlled contraception not as a burden but as a blessing.

Thus by 1953 all of the pieces were in place for the development of a oral contraceptive. Biologists had elucidated the physiology and endocrinology of reproduction. Chemists had figured out how to synthesize a more powerful analog of natural progesterone, and at least one company hoped to capitalize on the market. Pincus was both willing and eager to act as head scientist of the venture. McCormick contributed the necessary funds, and both she and Sanger provided motivation for the project to move forward. Within three years, *Science* magazine would publish the successful results of the research on the hormonal control of contraception.[51]

The Pill Project Begins

Soon after the meeting with Sanger in early 1951, Pincus initiated investigations into the effects of progesterone and synthetic progestins on ovulation and fertilization. He directed one of his staff scientists, Min-Chueh Chang, to design and perform a series of experiments using laboratory animals.[52] Chang administered the steroids to female rabbits by mouth, by injection, or by placement inside the vagina, and observed that both progesterone and progestins inhibited ovulation. The effects on fertilization were unclear. Chang and Pincus published these results in 1953, acknowledging the financial support of Planned Parenthood and two chemical companies (not Searle), which also supplied the steroids.[53] The next set of tests repeated the same experiments in rats, whose reproductive cycle is closer to that of humans. The scientists confirmed their earlier findings: both progesterone and synthetic progestins suppressed ovulation in rats as well as in rabbits. The rat experiments used compounds supplied by Ciba and Syntex, but did not include Djerassi's new 19-nor progestin. In the published paper, Pincus and Chang credited three sources of funding: the Dickinson Memorial Fund of Planned Parenthood (through which McCormick initially routed her contributions), the Rockefeller Foundation, and the Alden Trust, Worcester. The authors made clear the underlying motivations for this research: "In view of the medical problems of family planning and population control, investigation of the possibilities of administering chemical compounds to inhibit ovulation, fertilization, and implantation is of importance."[54] Amidst the prosaic discussion of the scientific article, Sanger's behind-the-scenes influence shone through.

By mid-1953, the project moved into high gear, spurred in large part by the munificence of Katherine McCormick. The third series of animal experiments at the Worcester Foundation tested the reproductive effects of dozens of synthetic progesterone-related compounds, which Pincus obtained from several pharmaceutical companies, including Searle and Syntex (via its American distributor, Chemical Specialties Company, Ltd.). Preliminary results indicated that the group known as 19-nor progestins exhibited the ability to inhibit ovulation, and further experiments identified Searle's norethynodrel and Syntex's norethindrone as the most promising antifertility compounds. Pincus reported these findings at the Fifth International Conference on Planned Parenthood in Tokyo in October 1955; the results appeared in print in the conference proceedings and in two scientific journals, *Endocrinology* and the widely read *Science*.[55]

In that same issue of *Science*, Pincus also published a report on the effects of the 19-nor steroids on the human menstrual cycle. As soon as he had learned that these compounds inhibited ovulation in rabbits and rats, he wanted to try them in humans. In order to do so, he had to enlist the help of a physician.

From Clinical Trials to FDA Approval

For clinical trials with progesterone and synthetic progestins, Pincus turned to John Rock, an obstetrician–gynecologist at Harvard Medical School. Rock was a valuable addition to the project team.[56] An observant Roman Catholic, he stood a better chance of convincing the Catholic laity and hierarchy of the merits of contraception than did other non-Catholic birth control advocates. Margaret Sanger, at first leery of Rock because of his religious background, soon appreciated his contribution to the pill project. She wrote in a 1960 letter, "Being a good R.C. and as handsome as a god, he can just get away with anything."[57] Second, Rock's faculty position at a prestigious medical school gave him the respectability that Pincus and his colleagues at the little-known Worcester Foundation lacked. Third, Rock's primary research interest centered on problems of infertility; he had become renowned for his efforts to help couples to have children. Rock was a relatively recent convert to the cause of population control. Only in the years after World War II did he become convinced that smaller families, made possible by contraception, were the necessary response to the threat of overpopulation.

In the early 1950s, independent of Pincus's work in Worcester, Rock successfully induced pregnancy in previously infertile women by treating them for several months with estrogen and progesterone. Although the steroids prevented pregnancy during the course of therapy, some of the women conceived when the treatment ended; this phenomenon became known as the "Rock rebound effect."[58] When Pincus learned of Rock's work, he asked the physician to join forces in the hunt for an ovulation inhibitor, and Rock agreed. Pincus suggested two changes in the experimental regimen: use only progesterone (estrogen promoted cancer in laboratory animals) and administer the hormone for twenty days each month (to allow for a period of menstruation). Rock achieved the same rate of success in curing infertility (about 15%), but a significant problem remained: tests indicated that about 15 percent of the women ovulated while taking the progesterone.[59] Pincus and Rock needed to find an orally active compound that would completely inhibit ovulation. It was time to test the 19-nor steroids in humans.

Pincus, Rock, and their associates conducted the first set of tests using norethindrone and norethynodrel on fifty infertility patients at the Fertility and Endocrine Clinic and the Reproductive Study Center at the Free Hospital for Women, where Rock worked, in 1954 and 1955.[60] The synthetic hormones successfully inhibited ovulation and at the same time enabled 14 percent of the volunteers to become pregnant after the course of treatment. In the published scientific paper, the authors emphasized the significance of the compounds in treating infertility and played down their contraceptive properties.[61]

Two more groups of people acted as subjects in tests of "the physiology of progesterone." Twenty-three female medical students at the University of Puerto Rico School of Medicine agreed to participate in clinical tests in the spring of

1955; however, by summertime half of them had dropped out of the study. In order to measure the effects of progestins on ovulation and menstruation, the subjects had to take daily vaginal smears and temperature measurements, 48-hour urine collections, and have monthly endometrial biopsies.[62] Perhaps the women also objected to the unfriendly attitude of the project's coordinator, who tried to intimidate the volunteers into cooperation. He notified Pincus that "if any medical student exhibited irresponsibility of this sort [collecting urine samples over 24, instead of 48 hours], I would hold it against her when considering grades."[63]

The second group of "volunteers" came from the Worcester State Hospital; they might have been more accustomed (if not more willing) to submitting to medical tests. Pincus obtained permission from the relatives of twelve psychotic women (and also sixteen psychotic men) to test the effects of norethynodrel and an estrogen-progesterone combination. The purpose of these trials was not to verify the contraceptive capacity of the progestins but rather to study their effects on the reproductive system. Despite their dubious ethics by today's standards, these tests formed the basis of one of the first reports on long-term administration of 19-nor steroids in humans.[64]

It is important to consider the oral contraceptive clinical trials in the context of scientific experimentation in the 1950s. Research using human subjects was much less regulated than it is today. For example, from the 1930s to the early 1970s, the U.S. Public Health Service followed a group of African-Americans with secondary syphilis in Tuskegee, Alabama. In their investigation of the long-term effects of syphilis, the researchers provided no treatment to the afflicted subjects. (In 1997, President Clinton offered a formal apology on behalf of the U.S. government to the victims of the Tuskegee study.)

Other drug trials found subjects eager to participate. In 1954, more than a million school-age children were volunteered by their parents to be "Polio Pioneers" in the clinical trials of the Salk polio vaccine. In the 1940s and 1950s, the success of the so-called wonder drugs encouraged faith in the products of medical research and in the investigators conducting that research. Medical research continued unchecked until the mid-1960s, when forces both internal and external to the medical profession began to call for more stringent controls on human experimentation. This regulation came well after the completion of Pincus's pill trials. The selection of subjects for those investigations in the 1950s was not at all inconsistent with contemporary clinical research practices.

By 1956, the pill project was poised to enter the next phase of development: large-scale field trials of the contraceptive effectiveness of the progestins. As with all new drugs, the pill was tested first in animals, then in small groups of volunteers under controlled conditions, and finally in large numbers of volunteers outside the clinical environment. Before moving into this final stage, Pincus and Rock had to make two decisions: which 19-nor steroid formulation to use and where to hold the trials. They chose to use Searle's norethynodrel because Syntex's norethindrone had demonstrated the possibility of slight masculinizing tendencies (based only on the animal data of enlarged testicles in male rats); of course, the long-standing relationship between Pincus and Searle surely played some role in the decision. Furthermore, Searle's compound benefited from the accidental contamination of norethynodrel with a tiny amount of estrogen (mestranol). This combination actually reduced breakthrough bleeding during the cycle of treatment, so Searle revised the manufacturing process to include the incorporation of mestranol (1.5%) into the 10-milligram norethynodrel tablet.[65] At the time, Pincus did not discuss the carcinogenic potential of the drug's estrogen component.

Laws against birth control precluded the possibility of trials in Massachusetts and Pincus and Rock decided to conduct the field trials in Puerto Rico. As a test site, Puerto Rico offered numerous advantages over other American locations. First, Pincus and Rock knew people at the University of Puerto Rico School of Medicine, and they felt confident that they could gather a reliable research team. Second, the island suffered from overpopulation, and in spite of the predominance of the Catholic religion, both officials and women wanted to control fertility. Third, the poorly educated population proved ideal for testing whether illiterate women could handle the cyclic regimen. In later years, feminists would criticize the Puerto Rican field trials, claiming, not that the trials represented medical paternalism or cultural imperialism toward the women of Puerto Rico, but rather that the trials did not adequately prove the long-term safety of the pill and thus jeopardized the health of both the Puerto Rican volunteers and all pill users in years to come.[66] However, these objections were not publicly voiced until a decade after the trials had taken place.

The story of the Puerto Rican trials has been told many times by both participants and historians.[67] The first series began in April 1956 with a hundred volunteers from a housing project in Rio Piedras, a suburb of San Juan, under the leadership of Edris Rice-Wray, a physician associated with the medical school, the Public Health Department, and the Family Planning Association. Within a few weeks, the researchers had to contend with negative publicity when the newspaper El Imparcial printed an article condemning the project and its director, Rice-Wray: "A woman dressed as a nurse and who alleges to be working for the state government, is distributing between the house wives some pills to avoid conception and to counteract the increase in population in Puerto Rico . . . The Secretary of Health believes that it is a 'bad combination' that the public officials of the State offer themselves to be used in a neomalthusian campaign, and he condemned also that the State Government is used as bait for contraceptive campaigns of private agencies."[68] Although this article scared some women away from the study, many more clamored to take their places, and the trials continued without further public incident.

By November 1958, there were 830 women who had taken the pill in supervised field trials in San Juan, as well as in Humacao, Puerto Rico, and Port-au-Prince, Haiti.[69] These women took the pill for as little as one cycle to a maximum of thirty-seven consecutive cycles. The trials demonstrated the success of norethynodrel as a contraceptive agent. The pill proved to be almost 100 percent effective in preventing conception; the researchers attributed the few pregnancies that did occur to participants' failure to take the pills on a consistent basis. Many women complained of side effects such as nausea, headaches, dizziness, and gastrointestinal discomfort. In her initial report on the Rio Piedras trial, Rice-Wray concluded that "it [Enovid] causes too many side reactions to be accepted generally."[70] Twenty-four years later, in an article written for a twenty-year retrospective on the pill, she reassessed her earlier judgment: "We could not have been more wrong!"[71] Indeed, although some of the Rio Piedras women dropped out of the study because of the side effects, many chose to continue to take the pill for the benefit of freedom from pregnancy.

By late 1959, Pincus, Rock, and the Searle management felt confident that Enovid had been sufficiently tested as a contraceptive.[72] The earliest reports from Rio Piedras had proven the contraceptive efficacy of Enovid; later trials confirmed these results. However, prior to the passage of the Kefauver–Harris Amendments to the Federal Food, Drug, and Cosmetic Act in 1962, the FDA approved drugs on the basis of safety, not efficacy.[73] The field studies indicated that Enovid did not adversely affect the reproductive system or the menstrual cycle, and that fertility returned within several months of discontinuing the treatment. Since the FDA had approved use of Enovid for gynecological disorders in 1957, Searle figured that it had cleared a significant hurdle in the path toward approval for contraceptive purposes.[74] However, the FDA had approved only the short-term use of Enovid, for periods of less than one year. If Enovid was to be prescribed to healthy women as a birth control agent, it might be used for twenty-five years or even longer. At issue was the long-term safety of Enovid, but no one seriously considered withholding the drug until decades of tests had been completed. In May 1960, the FDA granted approval for the prescription of Enovid for contraceptive purposes; the agency evaded the problem of long-term safety by approving use of the drug for no more than two years at a time. Of course, the FDA recognized the novelty of a drug intended for use by healthy women, but so long as Enovid met the criteria of safety, the government would not stand in its way.

The FDA that considered and approved Searle's application to market Enovid as a contraceptive in 1960 operated not only before the passage of the Kefauver–Harris Amendments, but also before the thalidomide tragedy became known. In 1961, German physicians observed that the mothers of babies born with malformed or missing limbs had taken the sedative thalidomide early in pregnancy. By 1962, thalidomide had caused thousands of birth defects in babies in Europe (the drug had not yet received FDA approval in the United States, and after the news from Europe never did).[75] Prior to the thalidomide disaster, Americans expressed optimism about the continual flow of new wonder drugs from the scientists' laboratories to the pharmacists' shelves. This climate of confidence and enthusiasm for the products of science and medicine favored the approval of the oral contraceptive in 1960. Certainly, this faith in science had informed both Sanger and McCormick's involvement in the pill project. It remained to be seen how physicians and women—neither of whom had played a significant role in the drive for a better contraceptive—would react to the availability of a birth control pill.

Chapter Two

Physicians, Patients, and the New Oral Contraceptives

Within five years of the FDA's decision to approve marketing of Enovid, the pill became the most popular form of birth control in the United States, prescribed by 95 percent of obstetricians and gynecologists. By 1965, six and a half million married women and hundreds of thousands of unmarried women had obtained prescriptions for oral contraceptives (the number of unmarried users cannot be determined accurately because they were not included in official reports). Among married women under the age of forty-five, 26 percent had used oral contraceptives since they had become available, another 19 percent thought they might use them in the future, and only 3 percent had never heard of the birth control pill.[1] Within certain subgroups of the married female population, the rapid adoption of the pill was even more dramatic: more than 80 percent of white, non-Catholic college graduates, aged twenty to twenty-four, had used oral contraceptives by 1965.[2]

Statistics from Planned Parenthood's annual reports confirm the swift and widespread acceptance of the pill in the early 1960s. In 1961, 14 percent of new patients at Planned Parenthood clinics chose to use oral contraceptives for birth control. Two years later, in 1963, the proportion of new patients choosing oral contraception was 42 percent; the following year, it jumped to 62 percent.[3] In other words, the percentage of women choosing the pill more than quadrupled within three years. By 1966, seven out of every ten women visiting Planned Parenthood clinics for contraceptive services walked away with a prescription for birth control pills in hand.[4]

Several factors contributed to the exponential growth in the number of oral contraceptive users from 1960 to 1965. First, physicians embraced the pill because it was convenient and easy to prescribe, and its prescription-only status increased their authority and control in the realm of family planning. Doctors also had a strong financial incentive to put their patients on the pill, because women had to revisit their office every six months to have their prescriptions renewed. Second, as will be seen later, Planned Parenthood's endorsement of oral contraception allowed its affiliate clinics to offer the pill to its clients at subsidized rates. By lowering the cost, Planned Parenthood enabled many women to use the pill who might otherwise not have been able to afford it. Third, newspapers and popular magazines helped to publicize the pill with their ample coverage of the new contraceptive method. As a result, both private and clinic patients asked their physicians for oral contraceptives. Thus enthusiasm on the part of both doctors and women for this new form of birth control helped it to achieve widespread popularity in just a few years.

Physicians and the Pill

Physicians in the 1950s and 1960s, like their counterparts today, most commonly learned of the existence of new drugs via the commercial channels of detailmen (sales representatives) and direct mail from the pharmaceutical company. Pharmaceutical companies sent advertisements and product literature directly to the physician; in addition, these drug houses sent out detailmen to meet face-to-face with doctors both to keep them abreast of the company's latest developments and to answer questions about the company's product line.

A sociological study in the 1950s found that three-quarters of the doctors surveyed named either detailmen or direct mail as their first source of information about a new drug.[5] When asked about the final source consulted before adopting a novel therapy, half of the doctors cited their colleagues or medical journals as most influential. It is interesting to note that this study considered journal advertisements to have negligible significance as either informing or legitimatizing sources. However, another study found that journal ads became important as a source of information on dosage and delivery when a physician decided to prescribe a new drug.[6] G. D. Searle and Company used journal advertisements, promotional literature, and detailmen in marketing Enovid. These commercial sources, combined with research papers published in medical journals and coverage provided by the popular press, gave physicians ample introduction to and information about the innovation of oral contraception.

After Enovid was put on the market in 1957 as a therapeutic agent for various gynecological disorders, there was almost immediately speculation in the medical community about its potential use for ovulation control. In July 1957, both Searle and Planned Parenthood considered it necessary to release statements to clarify the situation. Searle sent out letters to all of the nation's obstetricians, gynecologists, and general practitioners to "set the record straight" on the recommended clinical applications for Enovid.[7] However, the letters also noted that "there is adequate evidence to indicate that the drug will inhibit ovulation when the physician so chooses and that it is safe for this purpose in short-term medication" and described when to initiate therapy to achieve this effect. Although Enovid did not have formal endorsement for use as a contraceptive, a physician could prescribe it for such purposes. It is doubtful that many chose to do so; the point here is that the manufacturer notified doctors of the anovulant and therefore contraceptive properties of Enovid as early as 1957.

The chairmen of the medical and biologic research committees of Planned Parenthood issued a more circumspect statement. They acknowledged their interest in and support of studies on the potential use of synthetic steroids as oral contraceptives, but cautioned that "there remain many questions concerning possible side effects, particularly with long-term use" and concluded that "it is much too soon to regard this possibility [of Enovid as oral contraceptive] as assured."[8] Although Planned Parenthood physicians were advised not to use Enovid for contraceptive purposes, they were made aware of the prospect of such a clinical application in the future.

When Enovid was finally approved for use as an oral contraceptive in 1960, Searle sent its proselytizing corps of detailmen to inspire physicians to write more and more Enovid prescriptions. In the pages of its in-house newsletter, the company exhorted its salesmen to "weed out all the negative points and convince doctors to get patients started on Enovid TODAY . . . We are making each selected call with one objective: Enovid Prescriptions."[9] This Searle article advised the company representatives to avoid "unnecessary discussion" on the topics of cancer, nausea, and religion, and encouraged them to give doctors what they wanted to hear: "The Physician wants to be convinced that Enovid is . . . His drug; control of his patient on a month-to-month basis if he desires . . . Making the role of the physician assume greater importance in family planning."[10]

The importance of the approximately fifteen thousand detailmen in the United States should not be underestimated.[11] Most physicians during this period liked these company representatives and appreciated their roles in providing information on both new and established products, and the complimentary drug samples, gifts, and free lunches they proffered. Thus, when visited by one of Searle's men, a physician was likely to listen closely to the sales pitch for Enovid.

Although few physicians approved of the volume and content of promotional literature mailed to them by the drug companies, the majority opened their mail and gave the pharmaceutical publications at least a cursory review.[12] As noted above, direct mail often served as notice of the existence of a new drug product, but was rarely consulted by physicians as a source of further clinical or pharmacological information. The promotional literature on Enovid usually included a list of indications for the drug (not only ovulation control, but also dysmenorrhea, amenorrhea, premenstrual tension, infertility, endometriosis, recurrent and threatened abortion, and postponement of menstruation), dosages and prescribing instructions for each of these conditions, and possible side effects. The 1960 brochure, "Enovid for Long-Term Ovulation Control," labeled "For the Medical Profession Only," contained reports on clinical studies, toxicity studies, chemistry, human and animal pharmacology, and references, along with information on dosages and side effects.[13] By 1963, this booklet had expanded from twelve to twenty pages, and the number of references almost tripled.[14] For doctors looking for further information, these booklets provided useful starting points although they were biased toward the corporate interest.

In many ways, the advertisements for oral contraceptives in medical journals were merely condensed versions of the direct mail literature. Prior to 1960, Enovid had been advertised for the treatment of endometriosis, a gynecological disorder. The first advertisement to appear after the drug received approval for ovulation control did not promote it for contraceptive purposes, but instead emphasized the use of Enovid to postpone menstruation "for convenience, for peace of mind, for full efficiency on critical occasions."[15]

More than a year after the FDA approval, Searle finally advertised Enovid for use as a contraceptive.[16] One side of the two-page spread showed a drawing of the mythical persona Andromeda breaking free from manacles around her wrists, with the following caption: " . . . unfettered . . . From the beginning, woman has been a vassal to the temporal demands—and frequently the aberrations—of the cyclic mechanism of her reproductive system. Now, to a degree heretofore unknown, she is permitted normalization, enhancement, or suspension of cyclic function and procreative potential. This new physiologic control is symbolized in an illustration borrowed from ancient Greek mythology—Andromeda freed from her chains."[17]

This new therapy allowed physicians to liberate their women patients from the vagaries of the female reproductive system and to control it. Family planning had become a medical problem, to be treated pharmacologically by the physician. Previously, the doctor's role had been limited to fitting diaphragms and dispensing advice on rhythm and barrier methods; now, with Enovid available by prescription only, he moved onto center stage as a principal and essential actor in the control of fertility.

This initial advertisement, which appeared frequently in *Obstetrics and Gynecology* and in the *Journal of the American Medical Association*, described Enovid's mode of action as "induc[ing] a physiological state which simulates early pregnancy—except that there is no placenta or fetus . . . Enovid is as safe as the normal state of pregnancy." Safety became a key feature in the next advertising campaign, which showed on the left side an enlarged photograph of a tablet stamped "SEARLE" with the simple caption "the pill" and, on the right side, affirmative answers to the questions: "Is Enovid dependable? Is Enovid safe? Is Enovid practical?"[18] However, after reports linking Enovid to thromboembolism (a blood-clotting disorder) appeared in 1962, Searle dropped its confident assurance of the pill's safety in its printed ads and instead, noting Enovid's lack of androgenic properties, touted its product as "the first fully feminine molecule for cyclic control of ovulation."[19] (Julius Schmid Pharmaceuticals took advantage of the negative publicity surrounding the possible association between Enovid and thromboembolic disease in the summer of 1962 and promoted its Ramses diaphragm with the slogan, "No side effects with *this* method.")[20]

The Ortho Pharmaceutical Corporation also waited a full year after receiving FDA approval to advertise its birth control pill. In February 1963, the company advertised Ortho-Novum, its "new . . . well-tolerated specifically designed oral contraceptive," offered in a nifty package called the "DialPak" to help patients remember to take the pills regularly.[21] Now doctors had a choice of two anovulant steroids (Enovid contained norethynodrel, Ortho-Novum contained norethindrone).

By 1964, two more competitors had entered the field. Parke-Davis introduced Norlestrin with a three-page ad; Syntex's announcement of the availability of Norinyl occupied eight pages.[22] The potential market for oral contraceptives was vast, and the manufacturers spared no expense in advertising their products, not to the consumers (no prescription drugs were advertised directly to the public in the 1960s), but rather to the physicians responsible for choosing the preparations and writing the prescriptions. These ads emphasized the newly enlarged role of the physician in family planning: "With the use of Norinyl not only can the physician offer a dependable and physiologic approach to fertility timing that is under his control, but the opportunity to further his patients' marital happiness and to promote family harmony is greatly increased."[23]

By declaring that oral contraception was a major improvement in family planning from the woman's point of view, the manufacturers' advertisements suggested that physicians could expand their clientele by offering this popular method of birth control. Indeed, the financial incentive for physicians to incorporate the pill into their medical practices must not be overlooked.[24] The fine print on side effects and contraindications did little to detract from the positive images generated by these ads. Physicians may not have depended on advertisements for information on whether to prescribe a drug, but they certainly encountered many ads for oral contraceptives as they flipped through the pages of their medical journals, and the effect of this exposure could only reinforce their approval of the pill.

The authors of articles in medical journals wrote more cautious appraisals of the new contraceptive pills. Most of these were general overviews, reports on clinical trials, or pharmacological studies; not until 1964 did researchers begin to seriously examine the health effects associated with use of oral contraceptives. A 1961 article by Edward T. Tyler, director of the large-scale clinical trials at the Los Angeles Planned Parenthood Center, typified this early medical literature.[25] In this "special contribution," Tyler described the history of the development and use of synthetic hormones in gynecology and the effectiveness of oral contraception as established by several large clinical studies. He acknowledged that "all the answers are not yet in," and listed the concerns about possible long-term effects on the health of women and the children born to them. Tyler also outlined the main disadvantages of steroidal contraception: the high cost of the pills and temporary side effects (nausea, gastrointestinal disturbances, breakthrough bleeding, weight gain). This evaluation presented a relatively balanced view of the available facts and allowed readers to make up their own minds about the pill. Presumably, commentaries such as this one were discussed among physicians, who relied on one another's opinions and experiences in making decisions about new drugs and therapies. However, these sober, academic reviews of oral contraception had to compete with the attractive, upbeat advertisements run side by side in the same medical journals.

Clearly, physicians received ample notice of the oral contraceptive pill, and plenty of sources existed for those who sought further information. Of course, not all physicians zealously advocated the pill. Some objected to its use for religious reasons. Others questioned the wisdom of releasing such a potent drug for long-term use by healthy women, but their voices were drowned out by the chorus of acceptance. Also, other factors besides the knowledge of and enthusiasm for oral contraception among doctors influenced the incorporation of the pill into medical practice. One unusual circumstance was that patients knew about the new pill for birth control, requested prescriptions from their physicians, and thus played an active, participatory role in medical treatment. This development will be discussed more fully later in this chapter. Second, the pill, as a contraceptive drug, was caught up in the medical profession's doubts about its role in family planning. Physicians differed greatly in their attitudes and practices regarding contraception before 1960, and the new oral contraceptives accentuated the discrepancies in the delivery of family planning services by American physicians. The requirement of a prescription for oral contraceptives obliged physicians to confront the issue of birth control directly, as more and more of their patients came to them with specific requests for the pills to prevent pregnancy.

Planned Parenthood and the Pill

By contrast, the Planned Parenthood Federation unequivocally supported birth control. The organization's goal of voluntary family planning embraced any and all methods; their position stated that "*any* method of birth control is more effective than *no* method, and the most effective method is the one the couple will use with the greatest consistency."[26] Clearly, there was an advantage in having a wide variety of methods from which to choose, and Planned Parenthood viewed the oral contraceptive as a welcome addition to the armamentarium of contraceptive choices. Thus when the FDA approved Enovid, the national leadership of Planned Parenthood offered its blessing and endorsed the use of the pill in its affiliated centers.[27]

Initially Planned Parenthood expressed concern about the cost of the oral contraceptive pills. Within days of the FDA approval, Medical Director Mary S. Calderone wrote a letter to G. D. Searle & Company to inquire about the cost of Enovid, recognizing that "there will be many poor patients who will not be able to afford the drug in any way for whom we will have to raise funds to cover the cost."[28] By January 1961, Calderone had negotiated a special arrangement with Searle in which the company would sell directly to Planned Parenthood affiliates, thus bypassing the traditional channel of distribution via wholesale and retail druggists and cutting the cost of the pills by more than half.[29]

Despite the enthusiasm of the national headquarters, not all Planned Parenthood affiliates jumped quickly onto the Enovid bandwagon. In January 1961, less than one-third of the clinics offered oral contraceptives to their patients. Some of the delay may be attributed to administrative difficulties in procuring and distributing the pills, but many centers faced opposition from physicians and boards of directors. The reasons for this resistance are not clear; at any rate, it seems to have been short-lived, as 88 percent of centers reported that they offered the pill by the end of 1961.[30] Interestingly, Planned Parenthood never accounted for more than a small fraction of all birth control pill prescriptions in the United States. In 1969, only 4 percent of American women taking the pill received their prescriptions from a Planned Parenthood clinic.[31]

Planned Parenthood's national leaders were surprised and pleased by the widespread success of Enovid. They had calculated that the method would appeal "only to sophisticated users," but in fact, women of all educational backgrounds and income levels had no trouble taking the pills reliably.[32] Given the popularity of Enovid among its patients and the organization's commitment to offering a wide variety of contraceptive options, the Planned Parenthood Federation threw its weight behind the pill and joined the ranks of its most staunch advocates. However, it is important to understand that Planned Parenthood did not "push" the pill as a superior contraceptive method. In its posters, pamphlets, and television and radio spots, Planned Parenthood mentioned no single method by name, but rather emphasized the importance of family planning, by whatever method worked best for the individual. As it turned out, the majority of its patients chose to use oral contraceptives. These women, like most American women, first learned about the pill from the news media.

Early Media Coverage of the Pill

People outside the medical community did not have access to the same sources of information as physicians; they had to rely on the popular media for news of medical advances and innovations. Thus, for the general public, popular periodicals contributed greatly to the dissemination of information about oral contraception. Magazine articles offered the public its first glimpse of this new method of birth control and thus conditioned the initial response to oral contraception. News magazines, business magazines, those catering to the general reader and those directed to select audiences (women, men, blacks, consumers) all published articles on the pill. Similarly, the *New York Times* (and many other daily papers that took their cues from the *Times*) consistently reported news of the pill in particular and birth control in general. Americans received ample exposure to the latest scientific development that directly affected daily life.

In addition to the "facts" about the pill, the media also presented reports of potential side effects associated with its use. Most of these articles tended to be nonjudgmental, leaving the reader to make up her own mind. Yet many journalists gave the impression that "science" was working to improve oral contraceptives, and therefore current shortcomings would soon be remedied. The generally positive showcase for the pill in the pages of popular magazines could only help to enlarge the ranks of women choosing oral contraception.

The idea of oral contraception had been introduced to the American public prior to 1960. Throughout the 1950s, newspapers and magazines related news of research on the use of drugs to control fertility, as reported at scientific conferences and in journals such as *Science*. Researchers had explored and ultimately rejected phosphorylated hesperidin, lithospermum extract, and field pea extract as candidates for development into oral contraceptives. As early as May 1957, *Time* reported from the annual Searle stockholders' meeting that "some of the most hush-hush medical research has been pursued in dozens of laboratories in the effort to find a contraceptive pill," and that Searle was about to release a drug intended for menstrual disorders that also had anovulant properties.[33] In the next several months, a diverse selection of periodicals (*Time, Business Week, Consumer Reports, Fortune, Nation, Reader's Digest, True Story,* and the *New York Times*) reported on the clinical trials of Enovid in Puerto Rico. *Fortune*'s story on "The Birth-Control 'Pill'" in 1958 predicted that FDA approval was ten years away and thus concluded that "progestin pills hardly shape up yet as a vastly exploitable business."[34]

Articles written in the early 1960s, after the FDA approval of Enovid, often described the history of the research, development, and testing of oral contraceptives and attempted to present explanations of how the drug worked in the human body. Stories that recounted the history of the development of the pill portrayed Pincus and Rock as heroic figures: "the biologist and the physician joined forces" in the interest of progress.[35] Similarly, articles assured readers that the pill was "tested extensively" before being approved for sale, although they gave few details of the trials.[36]

Many reports offered a brief description of the pill's composition and mode of action. *Good Housekeeping* explained that "the synthetic hormones act very much like the natural hormones of the body that prevent eggs from being produced during pregnancy," thus implying that oral contraceptives merely imitated the body's natural processes.[37] Indeed, the Roman Catholic gynecologist John Rock took this position: "The pills prevent reproduction simply by modifying the time sequences in the body's own functions . . . when properly used for conception control, the pills serve as adjuncts to nature."[38] However, some reporters, in the journalistic tradition of covering all angles of the story, questioned the alleged harmlessness of oral contraceptives; one cautioned that "all hormonelike drugs are as powerful as TNT in the ways they affect much of the body's chemistry."[39] In general, journalists accurately described the female reproductive cycle and the effect of synthetic hormones, in varying degrees of detail. The action of the pill in the human body was neither mystified nor mythicized in the popular literature; however, the result of its use by women was subject to endless speculation about the impact on individuals, families, and society at large.

At first, discussion of the health consequences focused on the associated side effects. Most writers reported that some 20 percent of all women taking oral contraceptives experienced headache, breast tenderness, bloating, weight gain, dizziness, nausea, and breakthrough bleeding, but the figures given varied from as low as 5 percent up to 70 percent. Although they were presented as drawbacks, the gravity of side effects was minimized because both scientists and physicians considered the symptoms merely inconvenient annoyances that would disappear after a few months. In an article in *Ebony*, the president of Planned Parenthood noted that most pill-induced problems could be alleviated by medication. Thus, he recommended antacids for gastrointestinal disturbance, appetite depressants or diuretics for weight gain, and a doubling of the oral contraceptive dosage for breakthrough bleeding.[40] These prescriptions implied that medicine had an answer in tablet form for any minor discomfort associated with oral contraceptive use.

Journalists took the allegations that the pill could cause debilitating and perhaps even fatal diseases more seriously. Reports of women who had suffered from thrombophlebitis and thromboembolism while taking oral contraceptives emerged in 1961 and 1962, and received immediate attention by the press. Following so closely on the heels of the disclosure that thalidomide caused birth defects, the possibility of a link between the pill and blood-clotting disorders was not to be taken lightly. The *New York Times* reported almost daily on developments concerning the pill and thromboembolic disease both in the United States and abroad in mid-August 1962. The flurry of alarm died down after an investigation by the Food and Drug Administration failed to find any cause-and-effect relationship between Enovid and abnormal blood clotting, and the number of women taking the pill continued to increase in spite of the health scare.[41] A cartoon in *Playboy* summed up the public's nonchalant response to medical concerns about the pill in 1963. It showed a scantily clad cigar girl offering "cigars, cigarettes, pills" to a couple in a nightclub.[42] Still, concern was kept alive by the popular press and would be revived a few years later by studies published in medical journals.

In contrast to the documented cases of thromboses, there was no early indication of cancer associated with use of the pill. In fact, initial reports suggested that oral contraceptives might even *inhibit* the development of cancerous tumors in the cervix. In 1961 the American Cancer Society awarded a $58,000 grant to Gregory Pincus to study the anticancer properties of the pill.[43]

When an abstract in the *Journal of the American Medical Association* in 1964 reported that research at the University of Oregon showed that Enovid accelerated the growth of breast cancer in rats, journalists sided with doctors in rejecting the study as irrelevant to humans. The *New York Times* concluded that "informed opinion appears to hold that the pills are safe . . . most say that the data show that if anything they protect women against cancer."[44] *Time* blamed irresponsible journalism for blowing the study out of proportion: "any medical significance of their [the Oregon researchers] work has been overshadowed by screaming headlines . . . the consensus [among doctors at the AMA meeting] was more than reassuring: women who take oral contraceptives do not incur any added risk of cancer, said the experts, and there are even glimmers of hopeful though preliminary evidence that the pills may actually be protective against some forms of cancer."[45] Surely such a statement would have comforted women already taking the pill and perhaps encouraged others to choose it.

Although physicians were not fazed by the Oregon study, investors reacted, causing the price of Searle's stock to drop. *Newsweek* noted acerbically, "All the flurry seemed to prove was that when it comes to analyzing medical research, Wall Street is woefully inept."[46] Thus the public received the message that financiers and journalists ought not to be trusted on medical matters. Only scientific experts had the authority to judge the merit and significance of research, and they firmly rejected any causal link between the pill and cancer. Through the end of 1964, the media publicized the possibility that oral contraceptive use could protect against cancer; they dismissed concerns that the pill might promote cancer. This view matched the mood of optimism surrounding the pill; when popular opinion turned toward skepticism in the late 1960s, the relationship between the pill and cancer would be reexamined.

Beyond its effects on the personal health of a woman, the pill would also be judged for its potential impact on her morality. By and large, the use of oral contraception in the early 1960s was assumed to be limited to married women, so the controversy over the spread of the pill to college and high school students did not erupt until the middle of the decade. However, two early articles did address the issue, and they are notable first for their foreshadowing of the debate to come, and second for their moderation and dispassion toward the subject, especially when compared with the sensationalist discourse of later years.

In January 1961, *Mademoiselle* published an analysis of the results of a questionnaire sent to college women around the country that asked, "What, if any, effect will a completely reliable oral contraceptive—one that can be used safely by the experienced and inexperienced alike, that guards against the collapse of a well-planned future—have on the behavior of the college community?"[47] The author, a college junior herself, drew the following conclusions from the more than two hundred responses received: "First, that undoubtedly the number of girls who are not virgins at marriage will increase, but by too small a number to cause more than a ripple in our great ocean of sexual tradition; second, that the pill will have no effect whatsoever on most women's desire for sex with one man within a permanent love relationship. It is within marriage that the pill should have its greatest impact, making sex a happier, freer act for those who choose to limit their families . . . anyone who expects a moral revolution will almost certainly be disappointed."[48]

In other words, the pill was not expected to promote promiscuity, although increasingly liberal attitudes toward premarital sex, combined with the removal of the fear of pregnancy, would certainly influence some women to reject traditional morals and constraints on sexual behavior. The author reassured her largely teenaged audience (and perhaps their mothers also) that the birth control pill would not instigate radical social change but rather would lead to healthier, more satisfying relationships between mature individuals.

Gloria Steinem, in an article on sex and the single woman in the September 1962 issue of *Esquire,* took a different position in addressing her mostly male audience. She acknowledged sweeping changes in sexual attitudes and behavior, but did not wholly attribute them to the advent of the pill. Steinem wrote: "The pill is obviously important to the sexual and the contraceptive revolutions, but it is not the opening bombshell of either one . . . The fact that the contraceptive revolution is already in such an advanced stage may explain why the invention that marks its height and perhaps its completion—the first completely safe and foolproof contraceptive pill—is being accepted so quietly."[49] For this feminist, sexual freedom represented just one aspect of the liberation of women. She applauded the new breed of "autonomous girls" who, "like men, . . . are free to take sex, education, work and even marriage when and how they like" and questioned the theory that women's roles were biologically determined. The message to her male readers consisted of both a challenge and a warning. "The real danger of the contraceptive revolution," she concluded, "may be the acceleration of woman's role change without any corresponding change of man's attitude toward her role."[50]

Steinem's outspoken prescience was a rare exception in the early 1960s. Others may have been concerned about the impact of oral contraception on single women, but they did not discuss the topic in the pages of the popular press at this time. Instead, and in keeping with the assumption that only married couples practiced birth control, reporters directed their attention to the debate within the Roman Catholic Church over the morality of oral contraception as a method of family planning.[51]

For Roman Catholics, oral contraceptives added fuel to the fire that had raged for years over the admissibility of birth control. By 1960, the Roman Catholic Church remained the last major denomination to forbid its members to use "artificial" methods of contraception; only rhythm, the practice of abstinence during a woman's fertile period, was permitted. However, Catholic doctors felt increasing pressure to prescribe the pill for their patients, who begged to be allowed to use it for "therapeutic reasons," the euphemistic term employed by those who did not want to add to their already large families. One Catholic physician reportedly instructed his patients to take the pills for ten days a month, instead of the usual twenty, in order to regulate the menstrual cycle so they could more accurately determine which days were "safe" for sex.[52] He claimed that use of the pill on this restricted schedule did not inhibit ovulation, but merely facilitated the legitimate practice of the rhythm method. According to one commentator, "the debate [over the use of the pill by Catholics] moved gradually from the question, What are the circumstances in which the use of the pill would not be considered directly contraceptive and would therefore be permissible? to the broader question, Are there any circumstances in which the use of the pill, even when directly contraceptive, might be justified?"[53]

The most vocal Catholic advocate of the pill was John Rock. In 1963, he published a book entitled *The Time Has Come: A Catholic Doctor's Proposals to End the Battle over Birth Control,* in which he described the action of the synthetic hormones in the pill as simply an extension of the body's normal functioning. As "adjuncts to nature," the pills "merely offer to the human intellect the means to regulate ovulation harmlessly."[54] Although a *New York Times* reviewer dismissed Rock's rationale for the pill as "little short of preposterous" and "medical fantasy," he did commend the overall intent of the book.[55] Conservative theologians rushed to denounce Rock and his thesis and to reiterate the Church's position that artificial contraception was immoral. The debate fostered by *The Time Has Come* received wide publicity; the *New York Times* printed nine articles on the matter within the space of three weeks. Within a year, several magazines published major stories on Catholics and the pill, with Rock featured in each.

Rock himself wrote an article in the *Saturday Evening Post* that recapitulated his thesis, thus introducing his arguments to thousands of people who would not read the whole book.[56] *Newsweek* ran a cover story (with Rock as the cover photo) that explored the reasons for which the Church was supposedly reconsidering its position on birth control. Among those listed—ecumenical, economic, practical, political, humanitarian, psychological, and scientific—the last one was cited as the primary impetus for change. "[The pill] is the product of modern science, supposedly the adversary of religion; yet the pill opens up, rather than forecloses, opportunity for change within the church."[57]

This juxtaposition of science and religion must not be overlooked, for it represents an added dimension to the theological debate. An understanding of reproductive biology and endocrinology had become central to the analysis of the morality of oral contraception. Roman Catholics could no longer rely solely on the teachings of the Church to support their position on birth control; now, to enter into dialogue with advocates of the pill, they had to familiarize themselves with scientific facts and theories. Thus, *Look*'s article, "Catholics Take a New Look at the Pill," provided a lengthy description of the role of hormones in the female reproductive cycle to illustrate the mechanisms by which rhythm and oral contraception worked, so that readers would understand the issues at stake.[58] *Newsweek* declared, "Not since the Copernicans suggested in the sixteenth century that the sun was the center of the planetary system has the Roman Catholic Church found itself on such a perilous collision course with a new body of knowledge while all about swirl dangerous currents."[59] The authors of this article seemed to suggest that the optimism of John Rock was not unfounded, for the Church appeared to be headed in the direction of change.

Further evidence hinted that the time had indeed come for a change in Catholic doctrine. A Gallup poll taken in August 1962 revealed that a majority (56%) of Catholics thought birth control information should be available to anyone who wanted it.[60] When pollsters asked the same question in November 1964, 78 percent of Catholics concurred. Although the question concerned dissemination of contraceptive information, not approval of its practice, it is hard to imagine that three-fourths of all Catholics would condone the one without the other. Clearly, attitudes if not necessarily behavior were changing among the Catholic laity. When Gallup interviewers asked Catholics in July 1965 if they thought the Church would ever approve of birth control, 61 percent said yes. Thus, the stage was set for the announcement and reception of the papal encyclical *Humanae Vitae,* which addressed the subject of contraception in 1968.[61] In spite of the laity's approval of contraception, this document forbade the use of any form of artificial birth control. Lay Catholics were deeply disappointed by this conservative dictum, and many of them chose to ignore it. By 1970, two-thirds of all Catholic women and three-quarters of those under thirty were using birth control methods proscribed by the Church. Twenty-eight percent used the pill, compared with 33 percent of their non-Catholic counterparts.

For the Pope and his bishops, family morality lay at the heart of the matter. However, a larger issue loomed beyond the immediate family: the crisis of worldwide overpopulation. John Rock had declared that "the growth rate is as urgent a problem as nuclear energy."[62] In the early 1960s, Rock and others welcomed the development of oral contraception as a potential remedy to the so-called population explosion.

Whatever doubts may have existed about the pill's effects on individual health and morality, popular periodicals expressed enthusiasm for oral contraception as the first step toward an effective means of population control. These articles generally recognized that the pill did not provide the answer to the problem of overpopulation, primarily because of its high cost and its daily regimen, but they praised it for its effectiveness and its physical and temporal separation of contraception from the act of sexual intercourse. In July 1962, *Newsweek* featured the population explosion as its cover story.[63] While the article described the "much-publicized" pill as "the most promising technique for the immediate future" because of its effectiveness, it listed as drawbacks the cost, "human forgetfulness" in taking a pill a day for twenty days a month, side effects, and the advisability of taking hormones for an extended period of time. Similarly, the *New York Times* report on Ortho-Novum, the second oral contraceptive to be marketed, noted that it was "far from ideal" because its high cost placed it "beyond the reach of these millions in Asia, Latin America, and Africa who are most in need of family planning."[64]

Business Week, on the other hand, expressed much greater optimism about the worldwide prospects for the pill. Positively giddy over the financial success of Searle and the as-yet largely untapped worldwide market for oral contraceptives, *Business Week* editors believed that research would produce an ever more effective, lower-cost pill devoid of side effects. The only cloud on the otherwise clear horizon for the pharmaceutical industry was the possibility that the Soviet Union would produce a better pill than the United States, which would "raise the prospect of a possible East–West race, with political overtones, to supply cheap contraceptives to underdeveloped nations."[65] On the brighter side, the article cheerfully reported "an unanticipated side effect" of the pill: its link with lower rates of divorce, alcoholism, and crime, at least in the regions of Puerto Rico and Mexico where clinical trials took place. "Results from these test populations," *Business Week* added, "are leading scientists to believe that the security guaranteed by oral contraceptives solves many marital and psychological problems of poor and overpopulated families."[66]

Whereas cost presented a significant obstacle to distribution of oral contraceptives on a massive scale, poverty and illiteracy appeared to have little effect on the ability of women to use the pill effectively. A Kentucky physician reported oral contraceptives to be highly effective among a group of poor women in the South, for whom other methods had failed.[67] *Time* quoted Dr. Edris Rice-Wray on the appeal of the pill to women: "Even the poorest, with little or no schooling, are found to be faithful and conscientious users."[68] Here also readers may have received the message that they too were capable of using oral contraception. With family planning portrayed as beneficial, even vital, to the survival of humankind, American women may have found an additional motivation to use the pill besides their own personal convenience and desire for a small family.

In the early 1960s popular articles introduced the two promises of oral contraception, convenience for the individual and population control for society. Individual convenience was presented as tempered by the possibility of negative medical side effects, and newspapers and magazines devoted considerable space to these problems. However, they depicted population control as a universal good; journalists made no mention of the possibility that coercion and eugenics might rear their ugly heads. Concern about health hazards would escalate later in the 1960s, but questions about the propriety of fostering fertility control at home and abroad would not surface until many years later.

Much of the discourse about the pill in the popular literature was infused with a faith in science to continue to make improvements in its formulation. "Almost surely, better pills with fewer side effects are on the way."[69] "With the application of the method of science to the problem of human fertility, the development of adequate methods is inevitable."[70] In addition, popular articles on oral contraception stressed the importance of the role of the physician. The pill could be obtained only by prescription; thus, its use was "a medical judgment to be made by physicians."[71] *Good Housekeeping* declared its intent to provide its readers with "the latest authoritative medical information" about the pill.[72] These articles implied that while the drugs themselves might have disturbing side effects, scientists were working to correct these flaws and physicians had the wisdom to prescribe oral contraceptives prudently. In the early 1960s, the media depicted the pill in particular and science in general as holding tremendous promise and potential, and women leapt to avail themselves of the latest innovation of medical science.

Patients Request the Pill

A letter written in March 1961 by the executive director of the Denver Chapter of Planned Parenthood to Dr. Mary S. Calderone, the medical director of the Planned Parenthood Federation of America, reflected the popular demand for the pill among women and at the same time revealed the wide variation in prescribing practices among physicians: "Following the tremendous national and local publicity we are now being swamped with new patients requesting Enovid . . . We have several different clinicians working in our clinics, all with different ideas and feelings concerning the use of Enovid and different viewpoints on who is and who is not a good candidate for the use of Enovid. Naturally, many of our patients are friends and get together and compare notes and the results are quite catastrophic."[73] Both women and men wrote to Planned Parenthood or to Gregory Pincus at the Worcester Foundation to obtain more information about the new oral contraceptives and in some cases to request prescriptions for or direct delivery of birth control pills. Many of them described their predicament of too many unplanned children and looked to the pill for salvation:

> I am about 30 years old have 6 children, oldest little over 7, youngest a few days. My health dont seem to make it possible for me to go on this way. We have tried to be careful and tried this and that, but I get pregnant anyway. When I read this article [in *Science Digest* (September 1957)] I couldnt help but cry, for I thought there is my ray of hope.[74]

> Will you please send me some of these birth control pills. We have seven children we are both 30 year old. And we cant have children every year. So please send me these pills c.o.d.[75]

> I have been reading in one of the magazine books about your address of the birth control pills, and I wish you would help me very much. I sure need your help. You see sir I already have 9 children from age 13 to one years old, and about to have one in this days of August and I am only 29 years of age and I get very sick having babys one after another and I read about this birth control pill that could help someone like me, and make my life worth living.[76]

Correspondents often mentioned that they had read about the pill in a local newspaper or popular magazine: "Like thousands of others my wife and I bought a recent issue of *Coronet* because of the blurb on the cover concerning birth control pills."[77] Those who had read articles describing Pincus's work and the clinical trials offered themselves as guinea pigs for further research. Thus, a thirty-six-year-old woman with six children wrote to Pincus in 1957: "I read of your experiments which have resulted in a pill that will induce sterility . . . If it is at all possible, I would like to secure some of the pills, whatever the cost. I would also be more than glad to offer myself as a subject if you are still conducting experiments."[78] As discussed above, magazines and newspapers informed many people about the existence of the new oral contraceptives; those who did not actually see the articles themselves may have heard the news from friends or relatives who had read them. Women responded to the publicity by going to their doctors and asking for the pill.

The phenomenon of patient requests for the pill was unusual in the practice of medicine in the early 1960s. Traditionally, when a person felt ill and went to the doctor's office, he or she relied on the doctor to diagnose the condition and to determine the course of therapy. With the advent of oral contraception, that scenario changed dramatically. Women knew exactly what the problem was (they wanted to avoid pregnancy) and how to treat it (by taking the pill). All they needed from the physician was a written prescription to obtain the pills from the pharmacy. Previously, women had been shy about asking their doctors for contraceptive information and usually sought advice from nonmedical sources. Since oral contraceptives required a physician's approval, women were compelled to face up to their doctors to ask for the pill, about which they had heard so much.

Of course, many physicians gladly obliged by prescribing the pill. One doctor illustrated the optimism within the medical profession about both the safety and the popularity of oral contraceptives in his editorial comment on an article about the adverse health effects of the pill to be published in the *Medical Letter*, an independent noncommercial publication free of influence from pharmaceutical interests. The physician wrote: "This [report on the adverse health effects of the pill] is all more or less, mostly less, true. It's a typical sort of scare propaganda of half truths and facts taken out of context . . . Furthermore, it's like saying you shouldn't drive on the Merritt Parkway because you might get killed—certainly a true statement but what sense does it make. In short, I don't think publishing this would be worthy of the quality of the Medical Letter . . . P.S. I just bought some more Searle stock!"[79] The *Medical Letter* chose not to incorporate the reviewer's unbridled enthusiasm for the pill into the article, and instead concluded that "oral contraceptives should be considered only when topical methods do not serve."[80] The article seemed to have little influence on the acceptance of the pill among doctors and patients.

That direct requests from women accounted for much of the rapid popularity of the pill did not go unnoticed by the manufacturers. In the wake of publicity about the alleged link between Enovid and thromboembolic disease in 1962, Searle sent a letter to its sales managers and detailmen describing the negative effect on the public: "Any causal relationship between Enovid and thrombophlebitis is a far too complicated subject for the non-medical profession segment of the population to understand. They have been fooled badly. They have been scared badly. Many people believe that for a certain time period this will definitely slow down the number of requests by patients to physicians for Enovid therapy. As you well know, *this is our main and major source of increased and continued acceptance of the drug*" [emphasis added].[81]

Patients' requests for oral contraceptives affected not only the volume of sales but also the interaction between women and their doctors. Women who came to their physicians with specific requests for oral contraceptives no longer passively received medical care, but were transformed into active participants. In a sense, this circumstance increased the patient's power in her relationship with the physician, because if he refused to comply, chances were she would find another more willing physician. As one doctor commented, "You can't fight City Hall," meaning he could not turn back the tide of requests for the pill.[82]

This does not mean to imply that women had always been meek and passive as patients. There have been cycles in the history of medicine in which women have played more and less active roles in their treatment. In the midnineteenth century, women negotiated as equal partners with their physicians. Women dictated the circumstances of childbirth, including the decision to use anesthesia.[83] Females with consumption (tuberculosis) learned all they could about the disease and often challenged their doctor's opinions on treatment.[84]

The medicalization of women's reproductive health in the twentieth century interrupted the long-standing tradition of women as active participants in their health care.[85] In charting the course of American medicine in the period after World War II, in which physicians greatly expanded their authority and influence, historians have credited the new feminists of the late 1960s and early 1970s with challenging three decades of established social practices in the doctor's office and rejecting the imbalance of power in the doctor–patient relationship.[86] In fact, the initiation of such changes in the relationship between physicians and women can be traced to an even earlier date, namely, the early 1960s, when women sought the contraceptive services they desired.

There is an essential irony in the story of the pill's early popularity, which is that the pill served as a vehicle for greater authority in the field of contraception for both doctors and patients. This seemingly contradictory situation remained dormant in the early 1960s, when both groups expressed their satisfaction with this new method of birth control. Embedded in the heightened authority of both doctors and patients in contraceptive matters were the seeds of future conflict. Later in the decade, as questions about its safety came to light, women who had requested the pill felt confident enough to doubt their physicians' judgment and to demand full disclosure so that they could make their own informed decisions about using the pill. By the end of the 1960s, the medical profession, which had so easily stepped in to take control of family planning just a few years before, came under attack for perhaps overstepping its bounds. The development of the pill and its incorporation into American society took place during the heyday of medical science, but the changes it generated in the practice of medicine—namely, the incursion of physicians into the field of family planning and the empowerment of women in the doctor-patient relationship—would contribute to the controversy over the pill later in the decade.

Chapter Three

Sex, Population, and the Pill

The first decade of the birth control pill coincided with a period of monumental social upheaval in American society. From the civil rights movement to the student movement, from the New Left to the counterculture, from antiwar activism to women's liberation, the spirit of change was rife throughout the country. The changes sweeping the nation included the so-called sexual revolution, characterized by the liberalization of sexual attitudes, mores, and behaviors. Retrospective accounts of the 1960s often cite the pill as one of the causes of the sexual revolution. One researcher, for example, pointed out that the best alternative to the pill, the diaphragm, never achieved great popularity; while tolerable within the confines of marriage, it was "less suitable for the one-night stand." She doubted that the sexual revolution would have happened without the pill.[1] Other writers took the opposing view. They argued for sociological rather than technological causes of the sexual revolution, and dismissed the pill as either a necessary or sufficient antecedent.[2] The relationship between the sexual revolution and the contraceptive revolution, launched by the birth control pill in 1960, bears further investigation.

The twentieth-century trend toward sexual liberalism has been characterized by "the acceptance of sexual pleasure as a critical aspect of personal happiness."[3] Through the 1950s, Americans considered sexual liberalism to be acceptable within the confines of marriage, but this restriction could not stand up to the challenges of the 1960s. The swinging-singles culture that evolved was the product of not only the youth counterculture that developed on college campuses and in hippie communities but also of clever marketing targeted at young, unmarried, working individuals. Hugh Hefner's *Playboy* and Helen Gurley Brown's *Sex and the Single Girl* pioneered the cultivation and celebration of the single life.[4] Using these books as benchmarks, the sexual revolution would appear to be narrowly defined as an increase in sexual activity outside of marriage. However, it is important not to lose sight of its larger implication, namely, a greater openness about sex and sexuality in American society. This new frankness, manifest in myriad ways, was accessible to a larger segment of the American population. Not all college coeds "went all the way," but many traded in their girdles for miniskirts. A married mother of three may not have engaged in extramarital intercourse, but she could learn about the "myth of vaginal orgasm" in the best-selling book *Human Sexual Response*.[5] Most important for the purposes of this study, women in the 1960s could take the pill.

Of all the available methods of birth control that separated sexual intercourse from reproduction, the oral contraceptive prevented pregnancy most effectively. Women appreciated this reliability as well as the independence of this method from sexual intercourse. Perhaps the complete separation of contraception from intercourse was the pill's most innovative quality; women took their daily pills at a time and place unrelated to the act of coitus. Oral contraception was neither messy nor interruptive. Many women found swallowing a small tablet once a day for twenty days each month to be more appealing than fumbling with a diaphragm and jelly or persuading their partners to wear a condom. As Margaret Sanger had argued, women also preferred the pill because they controlled the method of birth control. A woman did not need the consent of her sexual partner to take the pill to prevent conception. Although later in the decade some feminists would become disillusioned with the pretense of sexual freedom, in the 1960s the pill offered many women the opportunity to enjoy sex with whomever they pleased without the fear of getting pregnant.

The aesthetic quality of the pill also helped to sanitize and to popularize sex. Its sobriquet, "the pill," offered a neutral term with which one could talk about birth control and by implication sexual intercourse. The acceptability of discussing sex in public paralleled the new openness regarding sexual matters in books, movies, advertising, and other aspects of popular culture. Magazines quickly heralded the new sexual revolution and speculated about its effects on society. For those who did not participate personally, media coverage provided the next best thing to being there. As Blanche Linden-Ward and Carol Hurd Green commented in their book, *American Women in the 1960s,* "Americans perceived social change as happening at a rapid and dangerous pace, while for the great majority, change came slowly if at all. Increased access to the news and pressure on news sources to fill their pages and broadcast hours meant that most Americans saw and heard much more than they experienced."[6] For example, *Time* illustrated a cover story in 1964 about "The Second Sexual Revolution" with photos of pornographic magazines, women in skimpy bikinis, and couples making out in public.[7] Such images could easily convince readers that the old rules governing sex and morality no longer applied in American society.

During the 1960s, journalists watched the innovation of oral contraception closely to determine its effects on sexual morality and activity. As time went on, they adopted the pill as the emblem of the sexual revolution. What is interesting about this representation is the implied relationship between the availability of the birth control pill and the shift toward more liberal sexual ethics. Articles in the popular media contributed to this association by speculating on the moral and social ramifications of easily obtainable, highly effective oral contraception. However, no data existed to confirm this hypothesis because no sociological research was undertaken in the 1960s to assess this relationship. Sociologists studied the incidence of premarital sex among college students, but did not correlate this activity with use of the pill (or any other contraceptive method). Demographic studies of the use of birth control included only married women; the first study of contraceptive use among unmarried teenagers was not conducted until the early 1970s. Researchers in sexology and family planning engaged in completely unrelated activities, producing noncomplementary data. The pill did indeed revolutionize birth control, and radical changes in sexual attitudes and conduct did take place, particularly among young people, but no one ever established a connection between these two phenomena.

Nonetheless, iconography and imagery of the pill as a symbol of the sexual revolution took hold and endured. In 1969, *Playboy* printed a cartoon of a beaming bride asking her new husband as they drove away from the church, "Can I stop taking the pill now?"[8] Whether or not the pill contributed to increased sexual activity among the unmarried youth of America, popular magazines fostered the perception of its link with liberalized sexual conduct, and this perception developed into a lasting impression. At the time, the pill figured prominently in discussions of changing sexual morality; thirty years later, the correlation between the availability of oral contraception and the sexual revolution persists. Demographic, sociological, and popular treatments of the pill, sex, and morality in the 1960s provide clues to the construction of the link between oral contraception and the sexual revolution.

Two important contradictions muddled contemporary evaluations of the pill. First, while the public generally accepted oral contraception as a superior form of fertility control, the use of the pill outside of marriage caused consternation among those who would preserve the sexual status quo. The conflation of the contraceptive revolution with the sexual revolution led many people to view the pill as "part of the problem" of sexual liberation among America's youth. Second, the proscription against use of the pill (or any other method of birth control) by unmarried women was generally limited to the middle and upper classes. On the other hand, birth control was recommended for poor women, both single and married, particularly those on welfare. These advocates considered the pill to be "part of the solution" to overpopulation.

However, not everyone agreed that birth control for the purpose of population control was beneficial. Some militant blacks portrayed the pill as a genocidal weapon, and they accused white population controllers of trying to diminish the black race in America. A minority of black men saw the birth control pill as another example of white coercion of blacks and fostered the image of the pill as a racist tool. Black women, on the other hand, rejected the notion of genocide as absurd.[9] When a mobile Planned Parenthood clinic in Pittsburgh closed after the leader of a militant group accused Planned Parenthood of discriminatory tactics and threatened to firebomb the office, the women in the community organized to seek the return of the clinic to their neighborhood. They resented the self-appointed spokesman for black Pittsburghers and argued that the use of birth control was a woman's personal decision.[10] Within parts of the black community, evaluations of the pill broke down along gender lines. For these women, as for so many other women around the country, the pill represented a measure of freedom: to choose how many children to have, if any at all, and when to have them.

These conflicting perceptions of the role and significance of the pill were colored by undertones of classism, racism, and gender bias. Concerns about the safety of the pill superseded consideration of its moral and political consequence in the late 1960s, but at middecade the public focused its attention squarely on the pill as an element of social change.

Sociological Research on Premarital Sex

One of the many indices used to gauge shifts in sexual standards measured the change in the proportion of women having sexual intercourse before marriage. Sex researchers relied on this statistic because it was quantifiable and thus lent itself to historical and cross-cultural comparisons. Unfortunately, emphasis on this statistic led to oversight of other important, albeit qualitative, transitions.

For example, reliable contraception (in the form of the pill) enabled many women to plan their family size, and this freedom helped transform their relationships with husbands, their involvement in activities outside the home, and their roles as wives and mothers. These effects would ultimately play a significant role in the evolution of American society, but being neither obvious nor directly measurable, they received little attention or publicity until the 1970s. During the 1960s, sociologists and journalists focused their attention on sexual attitudes and practices among teenagers and college students, and ignored the more subtle changes taking place among married women.

Many of these researchers set out to collect data on sexual behavior in the current generation in order to compare them with the results of Alfred Kinsey's landmark studies of 1948 and 1953. The first volume, *Sexual Behavior in the Human Male*, revealed that the majority of men experienced sexual intercourse before marriage.[11] Five years later, in *Sexual Behavior in the Human Female*, Kinsey reported that women born in the three decades after 1900 were much more likely to lose their virginity before marriage than were those born prior to 1900.[12] More than one-third of the younger cohort had engaged in premarital sex by the age of twenty-five, compared with one-seventh of the older generation. Although Kinsey tried to deemphasize the importance of this historical trend toward increasingly liberal female sexual behavior, others made much of this shift and proclaimed a sexual revolution among women in the 1920s.[13] Even so, rates of premarital intercourse for women remained far below those for men, and the "double standard," which condoned such behavior in men but condemned it in women, persisted into the 1950s.

Studies of premarital sexual behavior and attitudes in the 1960s fall into two categories. Those conducted prior to 1965 were intended to investigate sex in the post-Kinsey generation as follow-up studies to see what, if any, changes had taken place. In the latter half of the decade, there seemed to be a consensus that some sort of sea change in sexual morality had occurred, so that post-1965 studies assumed that change had taken place and set out to measure its nature and extent. Most of the research articles noted the popular conception of a sexual revolution and cited the extensive media coverage of this phenomenon as a stimulus for sociological investigation.

In endeavors to quantify the perceived trend toward liberalization of sexual attitudes and behavior on college campuses, a couple of research groups repeated their surveys of college students done ten years earlier.[14] In 1970, another group repeated its own 1965 study to characterize the pace of sexual liberalization as either evolutionary or revolutionary.[15] The editor of a 1966 issue of *The Journal of Social Issues* devoted to "The Sexual Renaissance in America," applauded the efforts of his fellow social scientists in the field of sexology: "Being bound by scientific demands for objectivity and impartiality, the social researcher provides a breath of fresh air in the hot and violent controversies that arise when matters of sex are discussed. The social scientist functions as a source of valid generalizations, of trend information, and of factual data."[16] He did not consider the possibility of investigator bias or methodological error in his laudatory portrayal of the contribution of social science.

Perhaps the most widely read of these sociological investigations was a book-length treatment by the popular author Vance Packard, who had written five other volumes of contemporary sociocultural criticism, including *The Status Seekers* and *The Naked Society*. In *The Sexual Wilderness*, published in 1968, Packard documented a wide range of changes in American society that he believed influenced the male–female relationship. They included demography, religion, fashion, music, cinema, television, and advertising. He preferred the term "sexual wilderness" to sexual revolution because "what in fact is occurring seems too chaotic and varied to describe yet as a revolution."[17] Packard did not limit his analysis to young people; he reviewed the broad expanse of relations between the sexes. To gather direct evidence on whether the subpopulation of college students responded to the changing morality by condoning or engaging in premarital intercourse, he initiated a survey of attitudes and behavior among students at several colleges in the United States and abroad.[18]

Relying on interviews and questionnaires for data, Packard's and the other studies of premarital sex suffered from methodological difficulties. First, they restricted their subjects to college students and therefore could not claim to be representative of the larger population of eighteen to twenty-two-year-olds. They also excluded those college graduates (and all others over the age of twenty-two) not yet married but nonetheless sexually active. Second, the sample size was quite small—usually a few hundred students and as few as forty-nine—and was often selected from a single university, so that even the college population was not necessarily adequately portrayed.[19] Third, there was no way to assess the candor of the respondents, especially those who completed anonymous questionnaires. Given the value-laden nature of the subject, it is not unrealistic to expect that some students may have been less than honest in their answers to such personal questions.

With these limitations, these studies can only provide at best an impressionistic picture of the sexual scene on college campuses in the 1960s. The earlier studies found that while there had been little change in the proportions of men and women engaging in premarital intercourse, students increasingly accepted the idea of sex before marriage. The later studies confirmed the shift in attitudes away from disapprobation and also detected an increase in premarital sexual activity among women. The percentage of men with premarital sexual experience remained constant; indeed, the figures had not changed since Kinsey's survey. More and more women, however, reported that they had sex before marriage, which led researchers to point to the erosion of the double standard. Not surprisingly, the sociologists had found what they were looking for: statistical evidence to support the contention that sexual standards were in flux.

Some of these researchers chose not to interpret their findings in a broader social context, relying instead on academic theory to evaluate the students' responses.[20] This analysis offered little insight into the relationships between changing sexual mores and contemporary social factors. On the other hand, Packard's sweeping portrayal of American society in the 1960s found little that did not have to do with the changing male-female relationship. Another research group identified the mid-1960s generation of college students as more rebellious than their predecessors: the students' involvement in the civil rights movement and the Vietnam War protests led them to criticize the norms of other social conventions, including marriage and sexuality.[21] However, the researchers gave little more than lip service to these social forces; they did not attempt to correlate student activism, for example, with approval of premarital sex.

Some sociologists, most notably John Gagnon and William Simon, both formerly associated with Kinsey's Institute for Sex Research, rejected the idea that any significant change had taken place in sexual behavior in the postwar era. In a scathing review of Packard's *The Sexual Wilderness* for the *American Sociological Review*, Simon dismissed the book's title claim: "One could just as easily have used these data to suggest not a wilderness but the presence of a formal garden slightly gone to seed."[22] Gagnon and Simon attributed the professional and public furor over sex to the novelty of sexuality as a topic for public consumption. People were thrust into a "new confrontation" with sex, manifest by extensive discussion about sex, particularly about people engaging in sex outside of marriage.[23] They claimed that the perception of an epidemic of sexual activity had developed, when in fact the only epidemic was one of talk and conjecture. Gagnon illustrated how easily the myth of a sexual revolution might be perpetuated. Upon prescribing contraceptives for a single woman, "a doctor with a dirty mind convinces himself that there is a sexual explosion. He talks to the press in the role of a marriage counselor, and then everyone else thinks so too."[24] Other sociologists agreed that the increased willingness to talk about sex led many people to believe that premarital intercourse was widely prevalent.[25] According to one, attitudes more than behavior had shifted in the direction of greater tolerance; whether behavioral changes would follow remained to be seen.

While sociologists did not agree on the nature and extent of the so-called sexual revolution among the youth of America, most (with a few exceptions, as noted earlier) recognized that the attitudes and behavior of college students in the late 1960s differed from those of their older brothers and sisters. Sex researchers employed a variety of terms in efforts to capture the essence of the perceived shift in sexual morality: "sexual wilderness," "sexual renaissance," "increased permissiveness." They often placed emphasis on the shift in female premarital sexual norms, but paid relatively little attention to the social forces contributing to these changes.

In particular, sociologists attributed a minor role, if any, to oral contraceptives in the development of new sexual standards among teenagers and college students. One group acknowledged the introduction of the birth control pill as a significant event in the 1960s, but noted that "the fear of pregnancy has not been a very important deterrent to premarital coitus for a number of years."[26] Packard listed six background forces contributing to the sexual wilderness, one of which he called "changes produced by the life-modifying sciences."[27] He included in this group the birth control pill, along with vulcanized rubber (for condoms and diaphragms), intrauterine devices (IUDs), surgical sterilization, abortion techniques, and improved nutrition. While Packard recognized the pill as a "historic breakthrough" in contraception, he also considered such forces as the baby boom, higher education, technological innovation, religious and philosophical outlook, and international affairs to be instrumental in affecting the relationships between men and women. An interesting sociological study of teenage sex and illegitimacy cited a number of factors, such as limited access to family planning programs, erratic sexual intercourse, and high contraceptive failure rates, which suggested that widespread and effective use of oral contraceptives among sexually active unmarried teenagers was quite rare as late as 1968.[28] Most of the other researchers simply ignored the availability of the pill (or any other method of birth control) in their studies of sexual attitudes and behavior among the unmarried.

This exclusion of oral contraceptives in sexology studies is puzzling and at the same time understandable when the demographic research on contraceptive use is examined. On the one hand, the results of quinquennial national surveys of family planning from 1955 to 1970 indicated radical changes in the pattern of contraceptive use among American women during the decade of the 1960s as the result of the technological innovation of oral contraception. Given this revolution in birth control, it is hard to understand how sex researchers could have disregarded the pill. On the other hand, the reason for omitting birth control from studies of unmarried sexual activity becomes obvious upon closer inspection of the demographic investigations, which included only married women. None of the national studies surveyed never-married women; the authors of the 1970 report were the first to acknowledge this omission (and implicitly that premarital sex did exist).[29] Not until the early 1970s was a major study of birth control among unwed teenagers undertaken; until that time, there were simply no data on the use of the pill by unmarried women. However, for the married population, a contraceptive revolution was documented in the 1960s and received ample publicity.

Demographic Research on Contraceptive Use

The Growth of American Families studies of 1955 and 1960 were designed to gather information on fertility and family planning among married couples in America.[30] The studies' organizers considered the use of contraception to be one of the determinants of family size, so they included questions about contraceptive practices, methods, and effectiveness in the interviews. The first study surveyed a national sample of white married women between the ages of eighteen and thirty-nine. The second study polled an expanded sample that included older white women, aged forty to forty-four, and a small proportion of nonwhite women. These two studies present a statistical portrait of contraceptive use in the immediate prepill era, although the utility of this picture is limited because of the exclusion of many women based on age, race, and marital status.

In 1955, 70 percent of white married women aged eighteen to thirty-nine had used some form of contraception at one time or another; by 1960, that number had risen to 81 percent, with another 6 percent expecting to do so in the future.[31] Women cited condoms, diaphragms, and rhythm as the most popular methods. In 1960, interviewers asked women whether they would use the soon-to-be-approved birth control pills. Opinion was divided equally within the total sample, with 42 percent saying they would and 42 percent saying they would not. Among those women already using contraception, 47 percent liked the idea of a pill, and if Catholics (most of whom relied on the Church-approved rhythm method of periodic abstinence) were excluded, then more than half of the women were receptive to oral contraception.[32]

The 1965 National Fertility Study demonstrated the dramatic effect of the availability of the pill on contraceptive choice among married women. Although this survey included many more nonwhite women and older women, for the purposes of comparison with the previous studies, most of the statistics pertained to white wives aged eighteen to thirty-nine. Within the five year period 1960–65, one-third of these women tried oral contraceptives. When asked to name the birth control method used most recently, 27 percent named the pill, 18 percent the condom, and 10 percent the diaphragm. Ten years earlier, 27 percent had named the condom and 25 percent had named the diaphragm as their most recent method. The percentages reporting use of all other methods, including rhythm, withdrawal, douche, and spermicidal jelly also dropped accordingly, as millions of women chose to use the pill for contraception.[33]

The 1970 National Fertility Study confirmed that the popularity of the pill was no fluke. The pill accounted for fully one-third (34%) of all current contraceptive use among all wives under the age of 45 (note that the 1970 sample was expanded to include blacks and older women). The next most popular method was surgical sterilization (16%). Only 14 percent reported use of the condom and a mere 6 percent relied on the diaphragm. Although the number of IUD users increased from 1965 to 1970, the IUD still only accounted for 7 percent of contraceptive practice.[34]

According to these national surveys of more than five thousand women, oral contraceptives were the predominant method of birth control among married Americans in the 1960s.[35] Both industry and private foundation estimates supported these figures. For example, Searle reported that the number of new Enovid prescriptions filled each year increased tenfold from 1960 to 1962 (191,000 in 1960 and 1,981,000 in 1962).[36] The Population Council calculated the number of pill users on an average day in America to be four million in January 1965, five million in January 1966, and five and a half million in June 1966.[37] The Food and Drug Administration's *Second Report on the Oral Contraceptives* used data from the National Prescription Audit and from the pharmaceutical manufacturers to estimate that eight and a half million American women used the pill each month in early 1969.[38]

This phenomenon can be attributed largely to the rapid acceptance and widespread popularity of the pill as the birth control method of choice among married women in the 1960s. How did this trend affect other aspects of social change under the wide umbrella of the sexual revolution? There were two noticeable influences by 1970: the embrace of birth control by Catholic women and the decline in the birth rate in the United States.

First, the trend toward greater acceptance of "artificial" methods of birth control (that is, anything other than periodic abstinence) among Catholics increased dramatically from 1955 to 1970. In 1955, only 30 percent of white Catholic wives reported use of a Church-proscribed contraceptive; by 1970, that number had jumped to 68 percent.[39] The pill was particularly instrumental in facilitating this trend during the 1960s, so that by 1970 more Catholic women relied on oral contraceptives than on any other method, including rhythm. The 1968 papal encyclical that reaffirmed the Church's stance against birth control had little, if any, effect on the laity. One woman commented, "If all the women who take the pill stopped going to church, there would hardly be any women there, only men and children."[40] By 1976, the disparity in pill use between Protestant and Catholic wives had disappeared; oral contraceptives were used by one-third of white married women aged fifteen to forty-four, regardless of religion.[41]

Second, the American birth rate began to decline in 1957, three years before the pill became available to the general public for contraceptive purposes. Although fertility continued to decrease during the 1960s, demographers were reluctant to assign any role to the pill, because historical evidence, such as the all-time low birth rate in the 1930s and the reversal of the postwar increase in the late 1950s, demonstrated that fertility need not be directly dependent on modern contraceptive technology. The authors of the 1965 National Fertility Study cautiously hypothesized that the pill might have influenced the degree, but not the direction, of the change in fertility.[42] Five years later, the correlation between effective contraception and the downward trend in fertility was more firmly established; the 1970 study considered the use of better methods of contraception "a major factor" in the declining birth rate.[43] Although the pill did not initiate this trend, researchers credited it with improving the ability of couples to space their children and plan their family size. One of the National Fertility Study directors further speculated: "An important by-product of postponing early childbearing is that it exposes the woman more to alternative interests potentially competitive with the mother role, such as working; this should also operate to reduce fertility further."[44] This comment, made in 1968, was prescient in its regard for the less obvious but potentially more significant transitions associated with the revolution in birth control.

As quantified by the Growth of American Families and National Fertility studies, the contraceptive revolution of the 1960s occurred among married (white) women. What about all those young women losing their virginity before marriage? Were they using birth control? This question can never be answered for the decade of the 1960s; only in 1971 did demographers first study contraceptive practices among unmarried women. This study found that 28 percent of more than four thousand never-married females aged fifteen to nineteen had engaged in sexual intercourse.[45] Although most had tried contraception, more than half had forgotten to do so on the most recent occasion and less than one-fifth reported consistent use.[46] The researchers speculated that the episodic nature of teenage sex might inhibit the use of contraception: "Family planning assumes both a family context and the possibility of rational planning. When, however, sexual encounters are episodic and, perhaps, unanticipated, passion is apt to triumph over reason."[47]

About one-fifth of the sexually experienced women reported using the pill as their most recent method of contraception.[48] The pill required a prescription, which meant a visit to the doctor or a clinic. Those girls with understanding mothers were lucky; the rest had to fend for themselves. Although by 1971, some physicians and clinics were willing to serve unmarried women, the majority still required proof of marriage. A dimestore wedding ring usually sufficed, but it might have required more nerve than most teenagers could muster. Some relied on older sisters or friends to obtain the pills, but this approach could result in a sporadic supply. In general, it was not easy for unmarried teens to acquire birth control pills and to take them consistently. More popular methods were the condom (used by 27%) and withdrawal (24%), both of which are nonmedical methods. Both methods are also the responsibility of the male; that is, more than half of the young women who reported using contraception at last intercourse relied on their partners for birth control. What is important for our purposes is to note the relatively small proportion of teenagers on the pill. Of course, the survey was subject to the usual problems of bias and interpretation of interview-based data. Furthermore, the sample consisted of teenagers only; the results might have been quite different had college students been the focus of study. Finally, the work was not published until 1973, long after the media had made the association between the pill and the sexual revolution and planted the images of this link in the public mind.

The Pill and the Sexual Revolution

Since demographers investigated fertility and family planning among married women, and sociologists studied the sexual ethics of unmarried people, it fell to journalists to correlate the contraceptive revolution with the sexual revolution. As discussed in the previous chapter, popular magazines and newspapers provided the primary sources of information for the general public about the pill when it was first marketed in the early 1960s. Continuing media coverage of its integration into the fabric of American life reflected its rapid acceptance. At the same time, newspaper and magazine articles helped to shape public opinion about the pill's social impact. In the mid-1960s, the pill figured prominently in articles on contraception, morality, teenagers, and family life. As journalists conveyed the impression of social upheaval to their readers, they implicated the pill as an influential factor.

What is interesting about the media's portrayal of the role of the pill in the sexual revolution is the variety of interpretations presented to the reading public. There was no consensus on the relative importance of oral contraception for modern morality, but rather, abundant speculation about the present and future consequences of the availability of the pill. In particular, articles tended to focus on the use of birth control pills by unmarried women; this phenomenon served as a touchstone for the expansion of sexual liberalization in America. At the extremes, depending on the editorial stance of the publication, single women on the pill spelled either progress or doom for contemporary civilization. Other writers took more moderate positions, including the pill as yet one more factor contributing to the new openness about sex. Some recognized that the availability of the pill compelled not only young women, but also the rest of society, to confront the role and regulation of sexuality in everyday life. In other words, the pill meant much more than just another method of birth control.

Controversy over the use of the pill by unmarried women erupted in October 1965 when newspapers reported that the director of Brown University's health service had prescribed oral contraceptives for two students. Prescribing contraceptives through the health service was seen to be inconsistent with other college regulations, such as curfews, designed to prevent students from having premarital sex. Students may have been engaging in such behavior, but university officials didn't want to know about it, much less do anything to condone or encourage it; in the mid-1960s the university still took seriously its *in loco parentis* role. On the other hand, underlying the general unwillingness at most colleges to allow contraceptive prescriptions to become part of the health services was a tacit acceptance of unmarried women obtaining birth control pills from private physicians under whatever guise necessary. The director of health services at Bennington said, "I do not feel it part of my duty either to pierce ears or to prescribe premarital contraceptives, devices, or medication."[49] Presumably the college allowed both earrings and birth control pills on campus, so long as the student procured them by her own means.

Journalists interpreted the Brown incident as symbolic of the intersection of the sexual and contraceptive revolutions, and used it as a starting point to comment on the social impact of the pill. An article in the *New York Times Magazine* considered the pill to be simply one more technological advance, akin to the television set and the automobile, to which Americans would make adjustments and eventually "take . . . for granted, and wonder how we ever lived without it."[50] According to this reading, the pill did not constitute a threat to society, but rather a challenge to reconsider existing standards of morality.

The conservative news weekly, *U.S. News & World Report,* took a much more alarmist stance. The subheadlines of its exposé on the pill posed four questions: "What is 'the pill' doing to the moral patterns of the nation? . . . Is the pill regarded as a license for promiscuity? Can its availability to all women of childbearing age lead to sexual anarchy? Are old fears of the social stigma of illegitimacy about to become a thing of the past?"[51] The authors of this article presented what they saw as disturbing evidence of a shift in sexual ethics away from restraint toward indulgence, and pointed to the pill as a causal agent. They sent a powerful message to their readers by making the case for a link between the advent of the pill and trends such as high school sex clubs and collegiate one-night stands. This construction portrayed the pill as part of the "problem" of premarital sex. Pearl S. Buck, the celebrated author, echoed this viewpoint in *Reader's Digest*: "Everyone knows what The Pill is. It is a small object—yet its potential effect upon our society may be even more devastating than the nuclear bomb."[52] Her atomic analogy would have been taken to heart in the Cold War years; comparing oral contraceptives with nuclear bombs labeled them as incendiary and dangerous. By appearing in this mass market publication, her words reached millions of readers.

Readers encountered more moderate interpretations in other widely circulated magazines. The theme of sexual ethics in flux was a popular one in the mid-1960s. To some writers, the pill represented another nail in the coffin of old-fashioned values. To others, it offered a welcome solution to the problem of unwed pregnancy, which was reported to be on the rise as a result of the sexual revolution. One obstetrician, writing under a pseudonym in *Ladies' Home Journal*, argued that in the absence of parental or societal guidance on matters of sexuality, physicians ought to prescribe birth control for unmarried girls: "A prescription for a contraceptive pill does not encourage immorality; it merely tries to cope with it."[53] While clearly disapproving of premarital sex, this doctor cared more about its unwanted results; he saw the pill as more of a solution than a problem in and of itself.

A buzzword that appeared consistently in articles on the pill and morality was "promiscuity." Laden with negative connotations, promiscuity was considered one of the main threats presented by oral contraception and was applied only to women. It had always been tacitly accepted that men would engage in premarital sex and that women, almost always referred to as girls, would remain virgins until marriage (it was never made clear with whom the boys were having sex). When it became obvious that "nice girls" were having sex (which, according to Kinsey, a good proportion of them had been doing for decades), a scapegoat was found in the pill. Furthermore, what constituted promiscuous behavior was never made explicit. The pill's detractors intimated that the pill might encourage women to "sleep around," that is, to have more than one sexual partner, because the fear of pregnancy had been eliminated. Pill advocates denied this association, as in *Time*'s cover story on the pill: "The consensus among physicians and sociologists is that a girl who is promiscuous on the pill would have been promiscuous without it."[54]

The mainstream press still condemned sexual activity by unmarried women, unless it took place in a truly premarital relationship (i.e., during engagement). Sexual liberalization had been narrowly extended for women to include sex with their husbands-to-be; anything more was regarded as promiscuous and therefore undesirable. Regardless of the article's orientation, a correlation between promiscuity and the pill was considered to be a bad thing. More conservative writers warned that widespread pill use among unmarried women would lead to what they saw as further dismantling of existing sexual ethics; their more liberal counterparts reassured readers that the pill would be used mainly by married women and would contribute to happier marriages and families. The point here is that both groups frowned upon unrestrained premarital sexual activity; they differed on whether the pill was part of this problem. The role played by the pill in furthering sexual liberalization or contributing to promiscuity depended on the editorial slant of the article; either way, the popular press implicated oral contraceptives to some extent in the sexual revolution.

The Pill and Population Control

The sexual revolution, as one *New York Times* writer observed, was mainly a middle-class phenomenon.[55] As such, it was the morals of middle-class teenagers and college students that were at stake in the age of the pill. On the other hand, family planners and population controllers wanted to provide poor women, married or not, with birth control, saying that the choice of family size ought not to be a restricted privilege of the middle and upper classes, but a right to be shared by everyone.[56] In 1965, the Office of Economic Opportunity (OEO) sought to justify the provision of birth control services to unmarried teenagers by speculating that well-to-do teenagers might have easier access to contraceptives, in which case the less affluent ought to have the same opportunity.[57] A year later OEO lifted its restriction on the use of funds to purchase contraceptives for unmarried women. Advocates of population control welcomed this news "from the viewpoint of national policy" as "a substantial breakthrough."[58]

Just six years earlier, President Dwight D. Eisenhower had repudiated the subject of birth control as inappropriate for government intervention. In 1959, he rejected the recommendation of the Draper Committee, an advisory group set up to review the U.S. foreign aid program, that assistance be offered to countries to help them "deal with the problem of rapid population growth."[59] He told reporters at a press conference: "I cannot imagine anything more emphatically a subject that is not a proper political or governmental activity or function or responsibility . . . This government will not, so long as I am here, have a positive political doctrine in its program that has to do with the problem of birth control.

That's not our business."[60] However, the president was clearly out of step with public opinion. In December 1959, a majority of Americans approved of making birth control information available both at home and abroad.[61] Institutional support for birth control also grew during this time; in the 1950s and 1960s, family planning organizations and population control groups forged an alliance to serve their mutual interests.[62] By 1963, Eisenhower the private citizen had publicly reversed his position on family planning, and in 1964 he agreed to serve as an honorary chairman of Planned Parenthood.[63]

The first real breakthrough in national policy on birth control came in January 1965, when President Lyndon B. Johnson announced in his State of the Union Address: "I will seek new ways to use our knowledge to help deal with the explosion in world population and the growing scarcity in world resources."[64] With Johnson's sanction of international population control came a concomitant interest in domestic family planning. Government officials realized that the United States should practice what it preached on the subject of birth control, so the federal government, mainly through the Department of Health, Education, and Welfare and the Office of Economic Opportunity's antipoverty program, became involved in sponsoring family planning programs around the country. At the same time, more and more states included family planning in their health services; from 1959 to 1966, the number of states operating contraceptive clinics rose from seven to thirty-five.[65]

The rationale for federal and state funding of birth control clinics was derived from the justification for committing taxpayers' money to help other countries deal with overpopulation. The international effort was initially undertaken as part of U.S. efforts to foster economic growth in underdeveloped nations. Foreign policy analysts considered rapid population growth an obstacle to social and economic prosperity. Policy-makers applied this philosophy to individual families (as opposed to the economy or society at large) in the United States also. They noticed that poor families tended to have lots of children, so they reasoned that they could help break the cycle of poverty by providing women with birth control. This simplistic solution was flawed in its narrow focus on fertility. It ignored critical factors such as lack of employment and lack of education that contributed to poverty. Nonetheless, federal agencies devoted significant funds and personnel to the development of contraceptive services in low-income areas.[66]

The judicial and legislative branches of the federal government also exhibited interest in family planning and population control. As indicated earlier, in 1965 the Supreme Court declared a Connecticut statute prohibiting the use of contraceptives unconstitutional. That year, members of the 89th Congress introduced several bills on birth control. The sponsor of one of the bills, Senator Ernest

Gruening of Alaska, held twenty-eight public hearings to obtain testimony in support of his plan to reorganize the departments of State and Health, Education, and Welfare to create positions responsible for coordinating the dissemination of birth control information.[67] Although the bill did not pass, both agencies increased their annual financial commitment to birth control during the mid-1960s.[68] Furthermore, the hearings served to increase the visibility of the perceived population problem and to focus congressional and public attention on family planning.

The full story of the U.S. government's participation in family planning and population control in the 1960s is beyond the scope of this work and has been chronicled elsewhere.[69] However, two relevant points deserve attention. First, the shift in government policy on birth control paralleled the liberalization of attitudes toward matters concerning sex in American society. Birth control was no longer a "private vice" but had become a "public virtue"; that is, contraception and family planning were now acceptable topics of conversation in mixed company and in public. A Planned Parenthood consultant reported in 1964 that "since January, 1963 . . . there has been a greater upsurge in the acceptance of public responsibility for family planning than in the whole history of the birth control movement."[70] A senior official of the U.S. Public Health Service told the president of the Population Council that "the interest in family planning is running higher all the time."[71] In Lyndon Johnson's Great Society, the federal government extended its reach into private life as it developed a broad range of social programs. Given the increased candor about sex and escalating concerns about overpopulation, the time was ripe for the government to move into the business of fertility control.

Second, technological innovations in birth control, namely, the pill and the IUD, facilitated the establishment of large-scale family planning programs by providing highly reliable, coitus-independent contraception. Both private and public birth control initiatives welcomed the pill as part of the solution. In October 1966, the Department of Defense included family planning services in medical benefits for military dependents. During the first nine months of this program, 98 percent of the almost half million women requesting birth control from military doctors received prescriptions for oral contraceptives.[72] Planned Parenthood also subscribed to an economic rationale for promoting birth control to the poor; as early as 1962, Planned Parenthood administrators calculated the costs of the use of the birth control pill versus pregnancy and child-raising among families on welfare and concluded that a midsized city would save about $75,000 in a five-year period if poor women gained access to the pill.[73]

Time magazine, in reporting a debate on reproductive choice at a Yale symposium in 1967, commented on the burgeoning interest in family planning in the age of the pill: "The open debate, covered matter-of-factly by the press, was further proof of the worldwide turnabout in attitudes toward birth control since the advent of oral contraceptives . . . In the past few weeks, newspapers and magazines have been filled with news of family planning, population control and the pill."[74] Given the popularity of the pill compared with other contraceptive methods among women, population control advocates believed that widespread distribution of oral contraceptives could help stem population growth.

The pill was not the cornerstone of all population control programs, particularly those targeted at underdeveloped countries. The Population Council primarily supported the IUD; more than 20 percent of the 393 grants funded by the organization in 1963–65 pertained to the development, testing, and implementation of IUDs for birth control, compared with less than 5 percent for the study of hormone-based contraception.[75] Decisions about the incorporation of birth control pills or IUDs into family planning programs seem to have been affected by the socioeconomic status of the prospective clientele. The president of Planned Parenthood, Alan Guttmacher, believed that private and clinic patients differed in their ability to use oral contraceptives successfully. Although the IUD had a higher failure rate than the pill, Guttmacher advocated IUDs in clinics and especially in underdeveloped nations. In a letter to the chairman of the board of G. D. Searle & Company, Guttmacher tried to convince him that IUDs would not compete with oral contraceptives:

> As I see it, the IUD's [sic] have special application to underdeveloped areas where two things are lacking: one, money and the other sustained motivation. No contraceptive could be cheaper, and also, once the damn thing is in the patient cannot change her mind. In fact, we can hope she will forget it's there and perhaps in several months wonder why she has not conceived.
>
> I do not believe the IUD's will cut into the competitive pill market materially in industrialized, more sophisticated regions. The big difference is that the IUD's are not as effective as the pill in preventing conception. If Mrs. Astorbilt, or Mrs. Searle or Mrs. Guttmacher gets pregnant while using an IUD, there is quite a stink—the thing is no good and a lot of people will hear about it. However if you reduce the birth rate of an unprotected segment of the Korean, Pakistanian [sic], or Indian population from 50 or 45 per 1,000 per year to 2, 3 or 5, this becomes an accomplishment to celebrate.[76]

Guttmacher did not confine his opinions to private correspondence. In public statements also he rationalized the differential use of IUDs and oral contraceptives: "For most private patients and more highly motivated clinic patients, the pill may give best protection. Where there is less motivation, the IUD is superior."[77] Although he left unanswered the questions of how to identify a "motivated" patient and who would make that determination, Guttmacher's insinuation that socioeconomic class and educational background were influential factors in contraceptive use smacked of classism. For Guttmacher and many of his colleagues in the field of population control, the pill represented the thinking woman's contraceptive, suitable for those patients conscientious enough to remember to take the pills consistently. The IUD was depicted as birth control for the masses.

While international population controllers championed the IUD as the easiest and cheapest method of birth control, that device never achieved the same popularity in the United States, where the majority of women who practiced contraception used the pill. There are several possible explanations for this difference: perhaps low-income women in the United States, like their more affluent neighbors, felt more comfortable with birth control pills, or doctors in family planning clinics found it easier to write pill prescriptions than to insert IUDs, or the pharmaceutical manufacturers pressured both private organizations and publicly supported clinics to buy the more profitable oral contraceptives instead of IUDs. In 1967, only 2 percent of American women used the IUD, compared with 30 percent using the pill.[78]

In the 1970s, after the health effects of the pill had been made public and feminists had seized upon the pill as an example of paternalistic health policies, concerns surfaced about inequalities in fertility control based on class and gender. However, in the mid-1960s, the plan to stimulate economic development abroad and to reduce poverty at home included reducing population growth by preventing women from becoming pregnant, as opposed to interfering with the male's role as impregnator, or more sweeping educational and social reforms. According to this policy, the pill was efficient, expedient, and emblematic of the technological potential to control human fertility.

By the late 1960s, the pill had become the icon of the sexual revolution, as evidenced by the photo on the cover of *Time* magazine in 1967 of dozens of birth control pills fashioned into the shape of the scientific symbol for "female." In 1968, Twentieth Century Fox released "Prudence and the Pill," Hollywood's first major motion picture starring the pill, along with David Niven and Deborah Kerr. Woven into the fabric of everyday life, the pill was firmly established in the popular culture. Few took seriously the conservative warnings that the pill would encourage young women in promiscuity. Although it is not known how many single women were using oral contraceptives in the 1960s, in the climate of greater openness about sex and sexuality, young people welcomed the possibility of hassle-free birth control. To the increasing number of young women attending college, joining the workforce, living on their own, and delaying marriage and motherhood, the pill presented an obvious advantage. By the end of the decade, casual acceptance characterized the attitude toward the pill among young people. In 1961, *Mademoiselle* ran an article on the moral and social implications of the new birth control pills; ten years later, in an article on the pill's effects on skin, hair, and health, the concern was largely cosmetic.

The mood of the 1960s fostered both individual use and widespread distribution of the pill for fertility control. People took to heart the message of population control both at home and abroad. In the early 1960s, economic rather than ecological discourse defined the problem of overpopulation; large families were the prerogative of the wealthy. The rhetoric of population control portrayed the pill as a weapon in the fight against poverty. Family planners recognized the same shortcomings of the pill as the women who took them, but they still regarded oral contraceptives, along with IUDs, as the best contraceptive methods for large-scale dispensation. In the climate of optimism that lingered into the mid-1960s, the pill epitomized the ability of science and technology to solve economic as well as social problems.

During this period, advocates of population control, sociologists, demographers, and journalists framed the effects of oral contraception in social, moral, or political terms. None of the voluminous discourse on the meaning of the pill detracted from its popularity and its integration into American culture. The merits of the pill came into question only when attention shifted to its effects on the health of its users. By the late 1960s, approval for the pill was tempered by reports of adverse health effects. Personal health concerns, not social or moral issues, ultimately set the terms of the national debate over oral contraceptives.

the first comprehensive regulator
of female cyclic function...

unfettered

From the beginning, woman has been a vassal to the temporal demands and frequently the aberrations of the cyclic mechanism of her reproductive system. Now, to a degree heretofore unknown, she is permitted normalization, enhancement, or suspension of cyclic function and procreative potential. This new physiologic control is symbolized in an illustration borrowed from ancient Greek mythology—Andromeda freed from her chains.

ENOVID *is an exclusive product of Searle Research. It is not available under any other name.*

F I G. 1. Although G. D. Searle & Company had advertised Enovid for the treatment of gynecological disorders since 1957, this was the first advertisement to tout the drug's contraceptive properties. (Advertisement in *Obstetrics & Gynecology* [July 1961], pp. 62–63.)

F I G. 2. To encourage physicians to prescribe Ortho-Novum, the Ortho Pharmaceutical Company advertised not only the specificity of its oral contraceptive but also the uniqueness of the pills' packaging. (Advertisement in *Journal of the American Medical Association* [February 16, 1963], page unknown.)

FIG. 3. Searle quickly met Ortho's challenge by designing the Compack tablet dispenser to compete with the Dialpak. (Advertisement in *Journal of the American Medical Association* [August 12, 1965], p. 193.)

"Cigars, cigarettes, birth-control pills . . ."

F I G. 4. *Playboy* promoted casual acceptance of the pill as early as 1963. (Reproduced by special permission of *Playboy* magazine. Copyright © 1963 by *Playboy*.)

"Can I stop taking the pill now?"

FIG. 5. *Playboy* (and *Newsweek*, which reprinted the cartoon) made it clear to readers that the pill enabled people to engage in sex before marriage. (Reproduced by special permission of *Playboy* magazine. Copyright © 1969 by *Playboy*.)

F I G. 6. "To Take or Not to Take." Scientific testimony at the Senate hearings only aggravated the dilemma the pill presented. (Reproduced from *The Washington Daily News* [Scripps Howard News Service], January 21, 1970.)

F I G. 7. Some Americans—in the wake of the Senate hearings—worried that abandonment of the pill would lead to unplanned pregnancies. (Reproduced from *Time* [March 9, 1970], p. 32.)

Chapter Four

Debating the Safety of the Pill

As the 1960s drew to a close, cynicism and doubt replaced the optimism and confidence that had pervaded America in the postwar era. Institutions that had been
venerated as pillars of modern society—the government, big business, universities, science, medicine—were scrutinized closely by members of the counterculture, journalists, and increasingly, the general public. A popular bumper sticker
exhorted its readers to "Question Authority." The pharmaceutical industry and
the medical profession fielded criticism about the necessity and wisdom of developing and employing so many expensive new drugs and medical technologies.
Writing in the 1970s, one researcher described the "pollution-conscious, post-
thalidomide sensibility, far different from the wonder drug optimism of the
1950s."[1] Dissatisfaction with the medical profession extended beyond charges of
excessive reliance on drugs and technology to a far-reaching indictment of the
practice of medicine. With rising costs, people began to resent the increasingly
impersonal nature of medical care and to question the expertise and judgment
of physicians. A historian of American medicine commented that "medicine, like
many other American institutions, suffered a stunning loss of confidence in the
1970s."[2]

The public lost confidence in oral contraceptives during this shift in attitudes
toward medicine in the late 1960s. After an enthusiastic reception and continued
popularity in spite of concerns about sexual morality in the mid-1960s, by the

end of the decade the birth control pill was surrounded by controversy. The debate centered not on its social or political impacts, but on how it affected the health of the women who used it.

All drugs carry some risk of side effects. Since most drugs approved by the U.S. Food and Drug Administration claimed to cure or relieve the symptoms of disease, decisions about the safety of using such medications usually lay in the hands of the medical experts, namely, the doctors. The pill, however, stood apart from all other drugs because it was taken by healthy women for long periods of time. When the press began to publish reports of adverse health effects associated with oral contraceptive use, the experts expressed a wide range of opinions about the relative risks and benefits of the pill, pregnancy, and other methods of birth control. As the controversy over the safety of the pill moved into the public domain, it created public interest in the evaluation of risk. In the absence of professional consensus, individuals sought information to help judge the risk of oral contraceptives for themselves.

The press played an important role in extending the controversy beyond the confines of the medical community to engage the attention of the general public. Earlier in the decade, the preoccupation of the popular media with the social and moral implications of oral contraceptives left little room for further speculation about their side effects. However, as the number of medical studies on their adverse health effects increased, media coverage of these reports escalated accordingly. By publicizing the results of medical studies on the health effects of the pill, the press heightened concerns among women about its safety. By presenting the contradictory nature of this medical discourse, the press also promoted doubts about the medical profession's ability to make wise judgments regarding the prescription of oral contraceptives. Since newspapers and magazines provided most people with their information about science, technology, and medicine, these articles played an important role in shaping public opinion about the pill. Just as women found out about the availability of the pill in the pages of popular periodicals in the early 1960s, so too they learned about the potential hazards of taking these pills later in the decade. Whereas the optimistic tone of those earlier articles persuaded many women to request the pill, the more skeptical tone of the later reports caused them to question the merit of oral contraceptives and the wisdom of their prescribing physicians.

The medical profession did not bear all of the blame for the pill crisis; after all, doctors also felt frustrated by the conflicting reports on the health effects of oral contraceptives and wanted the situation to be resolved. The pharmaceutical industry shared the call for accountability. It was no secret that the drug companies had made huge profits from the sale of oral contraceptives, and in the context of the broad-reaching social critique of the late 1960s, the motives and actions of the manufacturers did not go unquestioned.

Concern about the health effects of the pill paralleled other health-related issues that had captured the attention of the public in recent years. For example, the debates over the safety of radioactive fallout in the 1950s and pesticides in the 1960s weighed the alleged benefits of nuclear tests and pesticides against the potential risks to the health of exposed individuals. The controversy over the pill shared with these other debates the difficulty of assessing long-term health risks. However, a major difference distinguished the pill debate: individual choice. Whereas individuals had no choice in exposure to radioactive fallout or pesticides, women had a choice with the pill. In the final analysis, each person had to decide for herself whether the benefit of preventing pregnancy outweighed the risk of developing another condition. Public awareness of the potential problems associated with oral contraceptives did not make risk assessment or decision-making any easier for either physicians or patients, because those risks were not easily defined or measured. Patients desired more information so that they could participate in the decision whether to use oral contraceptives, but considering the inconclusive scientific evidence, physicians were unsure about what or how much information to give.

From Technical Debate to Public Controversy

The debate over the safety of the birth control pill can be characterized as a technical controversy, in which (1) the nucleus of the debate is a scientific or technological product or process, (2) some of the major participants qualify as expert scientists, and (3) experts on opposing sides dispute scientific arguments too complex for most laymen to comprehend.[3] The pill controversy met all three criteria. First, oral contraceptive drugs were the technological product of the application of scientific knowledge about endocrinology and steroid chemistry to the "problem" of fertility control. Second, well-respected physicians and clinical researchers sided both for and against the use of oral contraceptives. Third, these "experts" disagreed over the interpretation of statistical data generated by epidemiological studies and the significance of physiological effects of taking the pill. Both the math and the science involved in these issues were above the heads of most people.

The experts in a technical controversy take positions on the issues of need, effectiveness, and safety of the technology in question. Proponents of oral contraception argued for the pill's necessity on two counts: to provide women with a convenient, foolproof method of birth control, and to stem the tide of population growth. They pointed to the clinical success of oral contraceptives in the millions of women who used them as proof of the pill's effectiveness and safety. Opponents, on the other hand, downplayed the problem of overpopulation and emphasized the availability of safe and effective mechanical methods of contraception. They conceded that the pill had the highest degree of theoretical effectiveness, but claimed that in practice, when used properly, these other methods could be just as effective as the pill; moreover, errors of omission in the pill regimen seriously hampered its efficacy. The main thrust of the opponents' argument was that the pill could cause serious and sometimes fatal conditions in women. Of course, the controversy did not divide neatly into two equal and opposing sides. Within the medical community, there existed a continuum of support for the pill, ranging from complete and enthusiastic approval through varying degrees of acceptance for certain women under certain conditions to the extreme position in favor of a ban on the sale of oral contraceptives.

Had the technical debate over the effects of oral contraception remained within the purview of scientific research, it would have constituted little more than an interesting intellectual exercise, to be investigated with further clinical and laboratory studies. However, because millions of women took the pill each day, any information about its safety was of interest to the public, and the media made sure to bring the medical controversy to the attention of their readers. Although the pill debate did not develop into a mass protest movement, many Americans were aware of, interested in, and vocal about the pill controversy.[4] Women who took the pills began to express publicly their dissatisfaction with them, and physicians finally took heed of patients' complaints that they had previously dismissed. In 1969, the *New York Times* reported that "medical specialists at first tended to discount such unscientific comments [women's descriptions of physical discomfort], attributing them to the known vagaries of women. But now doctors are listening intently."[5] The idea that women had "known vagaries" would be attacked in the early 1970s by feminists seeking to empower women to regain control over their reproductive health; the immediate concern, however, of both women and physicians in the late 1960s was what to make of the conflicting medical reports about the safety of the pill.

Factors in the Development of the Medical Controversy

From the very beginning, scientists and laymen expressed concern that the use of oral contraceptives might be hazardous to a woman's health. Early clinical trials had revealed that women who took the pill experienced a variety of side effects, most commonly nausea, gastrointestinal disturbances, breast tenderness, weight gain, and breakthrough bleeding. Estimates of the percentage of women who experienced one or more side effects varied widely. One series of experiments demonstrated that women who took oral contraceptives and were not informed of possible side effects reported them much less frequently than women who were given the information and placebo pills in lieu of actual oral contraceptives.[6] Side effects bothered many women enough for them to stop using birth control pills in favor of alternative methods or no method at all.[7] However, physicians drew an arbitrary medical distinction between side effects and health risks that did not reflect patients' perceptions. Since doctors regarded many conditions as "minor" and "temporary" (regardless of how serious and long-lasting they might have been to women), these symptoms did not provoke significant controversy over the safety of the pill. Physicians did take notice of evidence suggesting that oral contraceptives might cause death or disease in the relatively healthy young women who took them.[8]

The medical debate over the safety of the pill took several years to materialize and once it developed did not yield easily to any sort of consensus. Three sets of conditions contributed to this situation. First, difficulties associated with data collection stemmed from the fact that the two major health concerns related to pill use—vascular effects (e.g., blood clots, strokes) and cancer (of the breast, cervix, and uterus)—occur very rarely in women of reproductive age, so very few cases appeared among women either on or off the pill. When reports of fatal blood clots in pill-takers were finally assembled, it was extremely difficult to ascertain the incidence of blood clots among the general population so it could be compared with the incidence among women taking oral contraceptives in order to assess whether there was a causal relationship.

Unlike thromboembolism, cancer does not arise suddenly but takes many years to develop into a pathological condition. The length of time between exposure to a carcinogen and the appearance of malignancy is called the latent period. In humans, most carcinogens (e.g., X-rays, radioactive paints, ultraviolet exposure) have a latency of at least ten years and usually more.[9] This delay makes it hard to ascribe the resulting effect to a certain cause; this difficulty is compounded by the fact that there are usually few symptoms or none during the period of latency. Thus in the 1960s, all discussion of the pill and cancer remained purely speculative because not enough time had elapsed for statistically significant numbers of malignancies to appear.

Epidemiological studies designed to trace the relationship between a possible causative agent (such as the pill) and the development of disease can take one of two forms, both of which have methodological drawbacks. *Prospective* studies follow two groups—for example, one taking oral contraceptives, the other using some other method of birth control—for several years to see what proportion of each group develops the disease. These studies are hindered by the large sample sizes needed and by dropouts from the program. *Retrospective* studies, also known as case-control studies, match a group with a disease with a control group that does not have it and compare the proportions of each group who used a drug—in this case oral contraceptives. These studies may be hampered by inappropriate selection of cases and controls, and by difficulties in obtaining medical (and here contraceptive) information about patients.

In addition to the general problems associated with epidemiological studies, there were some problems specific to the study of the health effects of oral contraceptives.[10] It was difficult to follow women taking oral contraceptives for a long time because many terminated use to become pregnant or to avoid undesirable side effects. Subjects moved away or failed to report for follow-up examinations. Furthermore, women who chose the pill for birth control formed a self-selected group, which meant that there might have been biases with respect to religion, socioeconomic class, race, and age, all of which can play roles in differential disease rates. Random assignment of subjects to different contraceptive methods would have been unacceptable. Also, physicians might have screened prospective oral contraceptive users according to their medical histories and physical examinations, so that those predisposed to certain conditions were advised to select an alternative method. Given these weaknesses in epidemiological studies of oral contraceptives, it is easy to see how the results of such research could be subject to multiple interpretations.

The second set of circumstances contributing to the complexity of the medical controversy about the pill consisted of problems associated with data interpretation and risk-benefit analysis. In the first years of oral contraception, the medical literature contained scattered reports of women who had become ill or died while taking oral contraceptives. The question that nobody could answer was whether the pill had caused the blood clots that afflicted these patients. Later in the decade, after several epidemiological studies on the relationship between pill use and blood-clotting disease had been completed, physicians continued to disagree on how to interpret these clinical research findings. Medical scientists disputed laboratory findings, such as the cytological classification of cervical smears on microscope slides in a study of cervical cancer among women taking the pill and women using diaphragms. When the data themselves were called into question, let alone the analysis of that information, there was little chance that consensus could be reached on the role of the pill in the development of acute or chronic disease.

Even those who agreed that oral contraceptives caused an increased risk of adverse health effects in the women who took them still disagreed on the significance of that risk. At issue was the complicated task of weighing risks and benefits for the individual, based on statistics derived from large populations. Herein lay the heart of the controversy among both the medical community and the general public. Different physicians (and patients) weighed the relative risks and benefits of oral contraception in different ways.

Central to this analysis was the perception of pregnancy as either a pathological or a normal condition.[11] If pregnancy was a disease to be avoided at all costs, then the benefits of the protection afforded by the pill might outweigh its attendant risks. On the other hand, if pregnancy was considered a normal stage in the female life cycle, then the pill not only suppressed that function but also carried with it unnecessary health risks. Even if the prevention of unwanted pregnancy was desirable, the availability of alternative methods of contraception added another consideration to the risk–benefit calculation.[12] For some, any risk of serious disease negated the benefits of contraceptive effectiveness; these opponents advocated the use of barrier methods of birth control. Proponents pointed to the higher failure rates of other contraceptive methods and to the higher mortality rates associated with pregnancy and illegal abortion to illustrate the relative safety and superiority of the oral contraceptive. In the case of cancer, the risk to pill users in the 1960s was unproven and remote.[13] The benefit of virtually foolproof contraception had to be weighed against the risk of developing a potentially terminal disease later in life. For a young woman concerned with the present threat of pregnancy, the risk of contracting cancer in thirty years' time as a result of oral contraception may have seemed too far in the future to worry about. On the other hand, physicians who had experience with the painful disease and death caused by cancer, when presented with the available, albeit circumstantial evidence, may have viewed the risks and benefits of oral contraception in a different light.

Regardless of whether they considered pregnancy as normal or pathological, most physicians and medical researchers referred to the pill as "oral contraceptive therapy." The use of the term "therapy" is curious, because there was no disease to treat in the healthy women who took the pill. What this term does indicate is the extent to which physicians had medicalized birth control. By referring to oral contraception as "therapy," doctors implied that the prevention of pregnancy was in effect a medical condition to be treated by medical professionals. The language used by doctors served to affirm their expansion into the realm of family planning.

The third determining factor in the evolution of the controversy over the pill lay within the medical profession itself, namely, its channels of communication. As a result of the increased specialization in medicine, doctors in different fields exchanged information less frequently, either informally by word of mouth, or formally in professional journals, because the different specialties did not often intermingle. The physicians who prescribed the oral contraceptives, mostly obstetrician-gynecologists and general practitioners, were not the ones who received and treated patients with symptoms of adverse health effects. Instead, a woman went to an ophthalmologist if she experienced vision problems, a psychiatrist if she felt depressed, or a neurologist if she had a stroke.[14] Ob-gyns rarely heard from patients with serious nonreproductive health problems, and specialists may not have thought to ask their patients about contraceptive use; thus, physicians did not quickly make the connection between oral contraceptives and adverse effects.

When reports of pill users with blood clots and other conditions were published, they often appeared in specialty or foreign medical journals that reached only small, specific audiences. The practicing physician, busy with patients, prescriptions, and paperwork, had time to skim only a few journals. Most general practitioners subscribed to the *Journal of the American Medical Association,* the *New England Journal of Medicine,* and perhaps the journals of their state medical societies. Obstetrician-gynecologists may have also subscribed to one or more of the journals devoted to their specialty. It is unlikely that these doctors would have seen the first report of blood clots in pill users, which appeared in the British journal *Lancet,* or the article describing eye and vision problems in pill users in the *Archives of Ophthalmology.* In fact, many doctors probably found out about these medical reports the same way their patients did: in the newspapers. As the number of articles on the health effects of the pill increased, as did the variety of conditions reported and the diversity of journals in which these reports were made, not even the most determined researcher could have kept up with more than a fraction of the literature. The combination of these three factors—poor communication within the medical community, problems of risk-benefit analysis, and difficulties in data collection—provided the framework for a controversy over the health effects of the pill that was at first slow to develop and then was resistant to resolution.

Medical Concerns and Media Attention

The first reports of fatalities associated with oral contraceptive use involved thromboembolic phenomena (inflammation of the veins, the formation of blood clots in veins or arteries, or obstruction of these vessels by blood clots) and appeared in Britain in 1961; by the following year, enough evidence had accumulated to prompt G. D. Searle & Company, the manufacturer of Enovid (the only brand on the market at the time), to sponsor a one-day conference at the American Medical Association headquarters in Chicago in September 1962.[15] Thirty experts met to analyze the data and to determine whether those data supported a causal relationship between Enovid and thromboembolism. The group consisted of physicians and scientists representing the fields of obstetrics and gynecology, medicine, surgery, biochemistry, pathology, and statistics, and included major luminaries of contraceptive research: Gregory Pincus, John Rock, Celso-Ramon Garcia, Mary Calderone (medical director of Planned Parenthood), and Christopher Tietze (research director of the National Committee on Maternal Health).

As of August 1, 1962, twenty-eight cases of death and disease from blood clots among the estimated one million or more users of Enovid had been reported. After the heightened publicity that month, 132 cases of blood clotting disorders in women taking Enovid were reported to the manufacturer by September 1, 1962. However, when cases with known predisposing causes were excluded, the number of cases reported among Enovid users was *lower* than that predicted for women not taking oral contraceptives. The diagnosis of thrombophlebitis or thromboembolism may or may not have been accurate in these cases; furthermore, cases that did not receive medical attention went unnoticed. There are several predisposing factors for thromboembolic disease (e.g., postoperative state, edema, obesity) that may or may not have been acknowledged by physicians; on the other hand, venous clotting can and does occur spontaneously, with no obvious physiological cause. While publicity about the possible link between Enovid and blood clots probably alerted most doctors to the importance and necessity of reporting cases of the disease in pill users, it may have deterred others who feared legal repercussions in the form of malpractice suits.

The conferees discussed the accuracy of diagnoses of thrombotic diseases, the validity of the available statistics, and proposals for studies to obtain definite and conclusive data on oral contraceptives and thromboembolism. When the chairman asked the group to vote on a resolution stating that the available evidence did not indicate a causal relationship between Enovid and thromboembolism, everyone agreed, except one person who felt the statistics were inadequate to exonerate Enovid as a thrombogenic agent and one who abstained from the vote. In addition, the group unanimously accepted a resolution calling for studies to gather additional information on the physiological and biochemical effects of Enovid, in both its contraceptive and noncontraceptive applications. John Rock advised that this second resolution be broadly inclusive in its wording to avoid "giving anyone the impression that we thought there was still a relationship between Enovid and thrombophlebitis which we have not yet been able to dig out, but if we keep on digging we think we will get to it."[16] Thus, the outcome of this conference was favorable for Enovid: "while more statistics are valuable, there is no real immediate danger involved."[17]

A conference sponsored by the manufacturer of the drug in question naturally evoked suspicions about impartiality. Furthermore, as publicity about blood clots in Enovid users increased, physicians reported more and more cases to both Searle and the FDA, so that by the end of 1962, the total had climbed to 272.[18] In January 1963, in response to growing concerns among the medical community about a possible link between the birth control pill and thromboembolism,[19] the commissioner of the FDA formed an ad hoc committee to determine "if the use of Enovid resulted in an increase in the incidence of death from thromboembolic conditions." Since the reporting of thromboembolic disease in both pill users and nonusers was extremely erratic, the nine-member committee of medical experts decided to restrict their data to deaths resulting from thromboembolism. They compared rates of mortality from thromboembolism in women taking Enovid and in the general population of similarly aged women, using case reports furnished by the FDA, Searle, and the National Center for Health Statistics. In the initial report, released in August 1963, the committee concluded that there was no statistical increase in deaths from thromboembolism among Enovid users under the age of thirty-five, but that inexplicably, those over thirty-five faced an elevated risk.[20] Five weeks later, the final report declared that as a result of corrections in statistical calculations, there was in fact no increased risk for any age group.[21] The conclusions of this group assembled by the federal government matched those of the group brought to Chicago by Searle the previous year: the available data did not support a cause-and-effect relationship between Enovid and thromboembolism.

Meanwhile, discussions in the early 1960s about the relationship between oral contraceptives and cancer considered the possibility that the pill might *prevent* cancer. In 1961, Gregory Pincus launched a large-scale study of the potential anticancer effect of oral contraceptives on the cervix that was funded by the American Cancer Society.[22] The final results of this research, presented in March 1965 in the journal *Metabolism*, revealed that the marginally lower occurrence of cervical carcinoma was not statistically significant; therefore the data provided no basis for assessing the incidence of cervical cancer among oral contraceptive users. Despite the inconclusive findings, the authors nonetheless interpreted them as suggesting that oral contraceptives had a possible preventive effect against the cancer.[23] The FDA's Task Force on Carcinogenic Potential rejected this interpretation and declared in 1966 that "there is no evidence to support the statement which has been made that the use of the contraceptive pills may have a protective effect against the development of malignancy of the uterine cervix."[24]

What was needed was a long-term prospective study; only after many more years would evidence be available to confirm or reject Pincus's hypothesis.

The first report to suggest a positive relationship between oral contraceptives and breast cancer was a seemingly inconspicuous abstract in the *Journal of the American Medical Association* in May 1964 on a report to be given at a session on malignancy during the annual AMA meeting. The abstract described a research project to determine whether a radioactive isotope of phosphorus in combination with steroid hormones would concentrate and curb the growth of tumors in rats. In the course of the study, the authors also found that the growth and development of preexisting breast tumors in rats was "markedly accelerated" by the administration of Enovid.[25] The news media picked up this report and gave the story far more publicity than might have been anticipated (see Chapter Two). In the actual presentation at the conference, the researchers mentioned acceleration of neoplastic growth by the estrogen-progestin compound (Enovid) only in passing and as a subject for further study. In the related article, published in the *Journal of the American Medical Association* in January 1965, the authors referred to a paper in press on the effect of these synthetic hormones on breast cancer in rats; however, that paper was not published until almost five years later.[26] At that time, the authors concluded that the oral contraceptive preparations were not inherently carcinogenic, but rather acted to enhance the growth of existing tumors in rats. They noted parallels between the experimental animal and human systems, but cautioned that the results could not be extrapolated directly to women.

Another animal study, this time involving dogs, focused the attention of both the medical community and the general public (thanks again to media coverage) on the possible link between oral contraceptives and breast cancer.[27] The pharmaceutical house of Merck, Sharpe & Dohme reported to the FDA that an experimental oral contraceptive containing a new synthetic progestin had produced breast tumors in four dogs given massive doses of the drug. In light of these findings, the company discontinued clinical studies already in progress.

Neither of the two reports on breast cancer represented traditional scientific research designed to test the relationship between the pill and cancer. The first, a study of the use of hormones to deliver radioactive isotopes to tumors, produced the findings on oral contraceptives as merely a by-product of the research. The second, a study of the effects of an oral contraceptive on tumorigenesis in animals, was conducted by the manufacturer to fulfill FDA requirements for new drug testing and did not have the results published in an academic or professional journal. In contrast to the growing body of empirical evidence of vascular problems in women taking oral contraceptives, there were no clinical data on the carcinogenic potential of oral contraceptives in humans.

Other clinical reports, this time documenting neurological and ocular disorders in pill users, were collected in response to an editorial in the October 1964 issue of *Archives of Ophthalmology*. The editorial specifically requested physicians to report cases in which patients using oral contraceptives experienced symptoms of strokes, migraines, or eye problems. The authors who initiated this investigation wanted to expedite the accumulation of information on any possible relationship between the pill and neuro-ophthalmologic problems: "It seemed reasonable that a single communication briefly indicating the need for combined effort might have a more immediate influence in directing attention to the problem, if indeed it exists, than would a series of individual case reports published in various journals at indefinite intervals."[28] The endeavor proved to be successful: within a year, the investigators had gathered more than sixty case reports from cooperating physicians, previously published accounts, and their own clinical experiences. In November 1965, they reported the results of their inquiry, taking care to emphasize that they intended to tabulate and summarize the collected information, not to draw premature conclusions about the existence of a causal relationship between the pill and neurological and/or ocular diseases. The authors did suggest, as did most investigators of the health effects of oral contraceptives, that the problem warranted further study.

Both the media and the FDA took notice of this report. For the first time, news of the pill made the front page of the *New York Times*.[29] The story caught the attention of the media and the public because the FDA asked the manufacturers of oral contraceptives to include in the list of contraindications to the prescribing physician a warning to discontinue use if patients developed eye problems, and to refrain from prescribing birth control pills to women with strokes in their medical histories. Although the link between oral contraceptives and neuro-ophthalmologic disease was not proven, the FDA considered the documentation of several dozen cases sufficiently important to advise that physicians and their patients be placed on alert.

Although news reports on the possible risk of eye damage associated with pill use appeared in newspapers and magazines, feature stories on the oral contraceptives continued to portray the pill as an instrument of social change, not as a potentially hazardous drug. In a lengthy cover story in the *Saturday Evening Post* on the birth control revolution that explored the meaning of the pill in the context of morality, religion, population control, law, and science, the author spent only a few paragraphs describing the "still unsettled questions about whether or not, in rare instances, the pills produce serious illnesses."[30] He mentioned cancer, thrombophlebitis, and eye problems as three possible effects of taking oral contraceptives, but emphasized that no causal relationship had yet been confirmed.

Similarly, a front-page article in the *New York Times* in 1966 assessing the impact of the pill on American life discussed its effects on family planning, marriage, sexual attitudes and behaviors, and morality before it addressed medical concerns. This author, like others, used the possibility of adverse health effects of the pill as a springboard for a discussion of "far more exciting . . . studies of new drugs that introduce entirely new principles of contraception."[31] These journalists conceded that the present pill had its faults, but they believed in the ability of medical researchers to come up with a newer, better method of birth control and thus maintained their faith and that of their readers in the miracles of modern science.

Other writers defended scientists and their work even more dogmatically. A hagiographic account in the *New York Times Magazine* of the roles of Chang, Pincus, and Rock in the development of the pill asserted that "criticism of the pill, limited in supporting evidence, has come in part from reputable physicians. On the other hand, many attacks stem from religious sources . . . Some opposition also comes from vested financial interests."[32] By linking medical concerns with venal businessmen and "rhythm cultists," as Rock called his religious antagonists, the author belittled those who questioned the safety of the pill. To allay any lingering fears, he quoted assurances from both Pincus and Rock that "all side effects . . . present figures of no statistical significance."[33]

Not all newspaper and magazine articles presented such sanguine portrayals of the oral contraceptives. In 1963, an article in *Redbook* cited concerns about the pill amidst contemporary misgivings about other technological "advances": "Controversy over Enovid would seem to be part of the increasing concern, among both doctors and laymen, that some scientists may be tampering dangerously with delicate balances of nature. The question 'Are we going too far?' reaches well beyond the use of drugs. It is reflected in public debate over radioactive fallout and over the use of pesticides that may be poisoning our food."[34]

Indeed, from the earliest days of the pill, some journalists (as well as physicians and women) stood apart from the general approval of the pill and questioned its safety. Morton Mintz, a reporter for the *Washington Post*, wrote several articles about the potential hazards of the pill. A seasoned veteran of the FDA beat in Washington, he had written about the thalidomide tragedy in 1962. Beginning in 1965, he criticized the FDA for its lax approach to clinical testing of oral contraceptives, its uncritical acceptance of the manufacturers' evidence, and its precipitous approval of Enovid in 1960.[35] Barbara Seaman, a journalist who wrote for *Brides'* and the *Ladies' Home Journal*, expressed her concerns about the pill's adverse health effects both in print and on radio and television talk shows.[36] Seaman took a different approach from that of Mintz; she wrote from the new perspective of women's health concerns. Both converted their research on the pill into book-length treatments at the end of the decade. Seaman was encouraged to write her book by Peter Wyden, the former executive editor of the *Ladies' Home Journal*, who had left the magazine to start a publishing company; it was perhaps this man who provoked *Time*'s assessment of the *Ladies' Home Journal* as "editorially allergic to the Pill."[37] However, without scientific evidence to support their qualms, these early critics made little headway in raising public awareness about the potential medical problems associated with the use of oral contraceptives.

By 1966, medical scientists had far more questions than answers about the health effects of oral contraception. Two comprehensive reports published that ·year indicated the extent of research activities on the pill and at the same time the dearth of definitive results. "Clinical Aspects of Oral Gestogens," released by the World Health Organization (WHO), summarized confirmed and postulated information about some thirty-odd effects of oral contraceptives on the systems and organs of the body. The findings of this report were inconclusive: oral contraceptives very effectively prevented conception, they produced "a number of deviations from established norms, but few, if any, of these appear to have pathological significance," and while serious adverse effects had been noted, "no cause-and-effect relationship has been established either by available statistics or by experimental evidence."[38] The committee listed twenty areas for further research on the effects of oral contraceptives and recommended that research be encouraged on the development of new methods of birth control as well. The report's encyclopedic content and neutral tone lent it an air of authority; however, the comment that "several controversial subjects were discussed . . . and Group agreement was not always unanimous" hinted at underlying discord.[39]

The report on oral contraceptives by the Advisory Committee on Obstetrics and Gynecology of the FDA resembled that of the WHO both in scope and findings. The group divided itself into four task forces, each charged with investigating one of the following areas: thromboembolic disease, carcinogenic potential, endocrine and metabolic effects, and efficacy. The committee recommended discontinuing time limitations for oral contraceptive use, since no scientific evidence indicated that prolonged use was more hazardous than short-term use. However, this suggestion seemed to contradict the group's additional recommendations, in line with those of their predecessors, to encourage and support further research. Specifically, they advocated a major epidemiological study of oral contraceptives and thromboembolism, and additional large-scale long-term studies of the effects of oral contraceptive use. The final conclusion was tentative: "The committee finds no adequate scientific data, at this time, proving these compounds unsafe for human use."[40]

The FDA's 1966 *Report on the Oral Contraceptives* included among its appendices a special report by one of the committee members, Roy Hertz, on three problems associated with pill use: carcinogenesis, thromboembolism, and the reproductive and genetic potential of ova exposed to oral contraceptives. Hertz's consideration of the cancer issue was reprinted almost verbatim later that year in the *Journal of the American Medical Association*; presumably, the publication of the paper in this periodical ensured that it would reach a much broader segment of the medical community. In case anyone had missed it, the article appeared a third time two years later in the *International Journal of Fertility*.[41]

Hertz's overriding concern centered on the unknown long-term effects of oral contraceptives. To his mind, the pill could not have been adequately tested, because not enough time had elapsed to reveal the results of taking oral contraceptives for twenty or thirty years. He argued that sufficient circumstantial evidence existed to support the hypothesis that the estrogenic component of the pill affected the course of breast tumor development: "Our inadequate knowledge concerning the relationship of estrogens to cancer in women is comparable with what was known about the association between lung cancer and cigarette smoking before extensive epidemiological studies delineated this overwhelmingly significant statistical relationship."[42] The other component of the pill, progesterone, had been shown to decrease the growth of endometrial carcinoma; Hertz pointed out that all cancer-destroying agents (e.g., X-rays) also induced cancer and therefore progesterone might also play a role in the development of cancer. Despite his concerns, Hertz did not call for a ban on oral contraceptives. Instead he advocated the continuation of long-term clinical studies, and he urged physicians to consider both the risks associated with oral contraceptives and the merits of alternative birth control methods.

In the wake of these two commissioned reports from the WHO and the FDA, the American Medical Association's Council on Drugs published its own evaluation in early 1967, based on the available scientific literature.[43] Since two-thirds of the physicians in the United States belonged to the AMA (and therefore received the organization's journal), many doctors read or heard about this latest assessment of the health effects of birth control pills.[44] In order to familiarize physicians with possible reactions to oral contraceptives, the article described a comprehensive list of side effects: breakthrough bleeding, failure of withdrawal bleeding, inhibition of lactation, breast tenderness, changes in thyroid and adrenal function, decreased glucose tolerance and worsening of diabetes, circulatory changes leading to blood clotting disorders, central nervous system and ocular problems, changes in liver function, edema, weight gain, nausea, and the controversial association with cancer. The Council recognized the challenge of considering the risks and benefits of oral contraceptive use for individual patients with only incomplete information at hand: "The development of the oral contraceptives...has broadened the meaning of the term therapeutics. For these pharmaceutic preparations are given to healthy people, and any possible hazards connected with their use, however remote, must be weighed against their expected benefits, perhaps more critically than has ever been the case before. This is often an uneasy responsibility for the physician, since the oral contraceptives are complex both in nature and actions and, as yet, these are poorly understood."[45]

This statement identified the crux of the debate over the oral contraceptives. What set "the pill" apart from other pills was that perfectly healthy women took it for long periods of time. All other medications were designed to cure an illness or to alleviate its symptoms. Vaccines, the only other drugs taken by healthy people, prevented the development of serious diseases. Oral contraceptives, however, were not technically "therapeutic"; their purpose was to prevent the condition of pregnancy. The physicians responsible for writing prescriptions for contraceptives viewed their wide use with mixed emotions, and as the list of adverse reactions grew longer, some became uncomfortable with their role in dispensing the pills. However, in the absence of conclusive evidence establishing a cause-and-effect relationship between oral contraceptives and severe adverse effects, neither practicing physicians nor government advisors wanted to pass judgment on the pill. Women had less access to these conflicting medical reports, and thus continued to take oral contraceptives unaware of the potential health implications.

When the FDA released its *Report on the Oral Contraceptives* in the summer of 1966, journalists made much of the lack of proof linking the pill to serious disease and the corollary conclusion that the pill was not unsafe. Articles titled "The Safe and Effective Pills," "Popular, Effective, Safe," and "Giving Pill a Safe Label" reinforced confidence in the pill as the best method of birth control available.[46] A *New York Times* editorial made clear both the importance and consequence of the pill's clean bill of health: "In the light of the world's great stake in checking the population explosion, the committee's certification that no conclusive evidence connects these oral contraceptives with observed ailments is doubly welcome. The result is likely to be an acceleration in the use of the pills and in the social revolutions they are promoting."[47]

Unlike their role in linking the pill to the sexual revolution, the media would not pass judgment on the health effects of oral contraceptives until they received the word from medical authorities. In the absence of any written medical report offering definitive evidence that oral contraceptives caused serious disease, journalists continued to minimize concern about their health hazards and to emphasize instead their social and moral impact. Thus, *Time's* big cover story on the pill in April 1967 announced: "In a mere six years it [the pill] has changed and liberated the sex and family life of a large and still growing segment of the U.S. population . . . Despite dark fears, there is not a shred of evidence that the pills cause cancer . . . Similarly, there is no evidence that the pills cause blood clots . . ."[48]

However, medical evidence was soon forthcoming. That month, the *British Medical Journal* published the preliminary results of epidemiological studies that linked the pill with a higher risk of blood clotting. Three months later, *Ladies' Home Journal* offered its more than six million readers an article entitled, "The Terrible Trouble with the Birth-Control Pills." This article ushered in a new era in reporting on oral contraceptives, an era marked by overriding concern about the health hazards of the pill.

In May 1967, the *British Medical Journal* published preliminary results of studies on morbidity and mortality from thromboembolism in oral contraceptive users, which was followed one year later by the final reports.[49] Using the case-control method, which matched groups of afflicted women with unafflicted women and compared the proportions of each group using oral contraceptives, the investigators found oral contraceptive users nine to ten times more likely than nonusers to suffer a venous blood clot and seven to eight times more likely to die from this condition. In absolute terms, 50 out of 100,000 oral contraceptive users would be hospitalized for thromboembolism each year compared with 5 out of 100,000 nonusers. The researchers also calculated the annual death rate from thromboembolism to be 1.5 per 100,000 oral contraceptive users aged twenty to thirty-four (compared with 0.2 per 100,000 nonusers in the same age group) and 3.9 per 100,000 users aged thirty-five to forty-four (compared with 0.5 per 100,000 nonusers).

The announcement of the British results raised the controversy over the safety of the pill to a whole new level. Previously, discussions of whether the pill caused blood clotting disorders had been handicapped by a lack of available information. Now, with statistics at hand, the medical profession disputed not only the interpretation of the data but the validity of the data itself. The completion at long last of epidemiological studies of the relationship between the pill and thromboembolism did not quell the medical controversy but rather served to stimulate vehement debate.

A few months later, a paper appeared in the *Journal of the American Medical Association* that presented contradictory evidence and conclusions, and flatly rejected both the methodology and the results of the British studies.[50] It is of interest that the primary author, Victor A. Drill, was affiliated with the Division of

Biological Research at G. D. Searle & Co. Drill did not conduct a comparable epidemiological study; instead, he reviewed the results of previous studies (such as the clinical trials of oral contraceptives) and found that the incidence of thrombophlebitis was lower in women using oral contraceptives than in nonusers. Drill reinterpreted some of the results of the British studies to support his own, opposing conclusions; those British data that did not mesh with his interpretation he dismissed as the result of flawed experimental design.

This article provoked a volley of letters to the editor of the journal over the next several months in which the authors of the British studies attempted to defend their research, Drill replied with his defense, and other readers criticized various facets of both studies. Medical scientists contested a number of issues during this exchange. First, the merit of the different methodologies was called into question. Whereas Drill objected to the retrospective method of study because the relatively small number of cases (only fifty-eight in the morbidity study) may have produced a large sampling error, the British researchers criticized Drill for relying on unpublished and incompletely published work.[51] Second, parts of the experimental procedure itself were challenged. One critic argued that the choice of cases for the British studies was flawed; by eliminating unmarried and pregnant women in addition to those with predisposing conditions, the British researchers may have skewed their results.[52] Third, publicity about the possible link between the pill and blood clotting was blamed for predisposing physicians to diagnose thromboembolism in oral contraceptive users more readily than in nonusers, thus biasing the findings.[53] The large number of variables present in the study of the relationship between the pill and thromboembolism made it easy to criticize the experimental design and to dispute the findings of such research.

The FDA's second report on oral contraceptives, released in August 1969, contained the results of a retrospective study of hospital admissions for thromboembolism among women of childbearing age in five American cities.[54] Similar in experimental design to its British counterpart, this study (named for its principal investigator, Philip Sartwell) used a much larger sample size. Unlike the British study and contrary to its original plan, the Sartwell study included unmarried women, because "it had become evident that idiopathic thromboembolism was relatively frequent among younger, unmarried women and it seemed unwise to disregard this source of case material."[55] By the late 1960s, it had also become evident that younger, unmarried women were having sex and were using birth control pills, and were therefore subject to the same adverse reactions as married women.

Sartwell and his colleagues found women who used oral contraceptives four and a half times more likely than nonusers to develop thromboembolism. Although these results corroborated those of the British studies, discrepancies in the data prevented this report from being the last word on the matter of the pill and thromboembolism. For example, the relative risk for American oral contraceptive users compared with nonusers was half as great as that for the British. Among the American cities, the relative risk of developing a blood clot for pill users compared with nonusers varied from 2.2 in Baltimore and Washington to 17.0 in Philadelphia. How could such wide variation be explained? Did Philadelphians on the pill really run such an increased risk of thromboembolism and if so, could they reduce that risk by moving to Baltimore? The Sartwell study was open to the same criticism as the British studies, namely, that the interview-based investigation was vulnerable to bias and that there were too many variables to control in the study of a disease that occurred relatively infrequently in women of childbearing age. Although many members of the medical profession accepted these studies of the positive relationship between oral contraceptives and thromboembolism, critics maintained that the findings could be interpreted differently.

Newspaper headlines and article titles from this period give some indication of the new critical focus and outlook of reporting on the pill. In 1969, the *New York Times* ran a front-page article on the "national impact" of oral contraceptives that focused exclusively on the health effects of the pill. Unlike the front-page review three years earlier, this article made no mention of morality, social impact, religious opposition, or population control. Readers of popular magazines also encountered "The Search for a Birth Control Method to Replace the Pill," "The Pill and Strokes," "Perils of the Pill," and "Doubts about the Pill" in the pages of their favorite periodicals.[56] An effective journalistic tactic consisted of presenting testimonials from women who had suffered ill effects from taking birth control pills. Reporters easily found women eager to tell their stories; these accounts often portrayed the physician as insensitive or negligent. Some articles explicitly blamed doctors for being remiss in the treatment of their patients: "The trouble lies not so much in the pills or their makers and takers as with doctors who prescribe them without doing a thorough physical examination and getting a good case history."[57]

Journalists also made much of the controversial nature of the debate over the health effects of oral contraceptives, noting that the experts differed in their opinions of the oral contraceptives. One science and health reporter observed that medical specialists, such as oncologists, tended to be more skeptical than primary care physicians, such as obstetricians, who prescribed oral contraceptives for patients.[58] Others used doctors as expert critics of the pill; by quoting physicians with doubts about the safety of oral contraceptives, a journalist could boost the credibility of his or her story. Such articles expressed confidence in physicians as medical authorities whose judgment deserved to be trusted. This approach fault-ed the women who took the pills for not heeding the caution dispensed by their physicians. The *Ladies' Home Journal* accused Americans of being "a people of enormous enthusiasm for 'the easy way.' Almost automatically we fall in love with anything 'modern,' and we are particularly quick to seize on new scientific 'mir-acles.'"[59]

By the late 1960s, the press had moved away from its prior fascination with the pill as a technological solution to the social problems of family planning and pop-ulation control. Morton Mintz, the *Washington Post* reporter, wrote that "a pub-lic predisposed to believe almost blindly in science and technology, deeply trou-bled by the desperate problem of overpopulation . . . such a public is not disposed to skepticism."[60] He argued that the public had been seduced by a range of au-thorities—medical researchers, physicians, manufacturers, and the press—into believing that the pill was the ideal "technofix" to solve both individual and so-cietal problems of fertility control, without being alerted to the potential health hazards accompanying the use of the drug. Mintz also charged that both medical and popular reports on the risk–benefit analysis of the safety of oral contracep-tives were disingenuous and therefore misleading. For example, Malcolm Potts of the International Planned Parenthood Federation made the following oft-quoted remark: "It would be more justifiable to have oral contraceptives in slot machines and restrict the sale of cigarettes to a medical prescription."[61] Mintz ar-gued that it was specious to compare the risk of dying from taking the pill to the risk of dying from smoking or dying in an automobile accident, because the be-haviors were not analogous. He also protested the practice of comparing the mor-tality statistics of pill users with those of pregnant women, because women who did not take the pill could use other, highly effective methods of birth control to avoid pregnancy.

An editorial in the *New York Times* in 1969 observed that "this hazard for healthy women taking oral contraceptives is different from the risk a sick person takes when receiving penicillin or a blood transfusion."[62] According to this rea-soning, oral contraceptives were qualitatively different from other drugs because they did not treat disease and because alternative contraceptive methods existed; therefore the argument that no drug, not even aspirin, was 100 percent safe did not apply to the pill. Nevertheless, magazine and newspaper writers used statis-tical information to offer purportedly scientific evidence about the relationship between oral contraceptives and thromboembolism. These statistics, combined with the personal testimony from patients or physicians in articles about the po-tential health effects of taking oral contraceptives, scared many women who may not have experienced any ill effects themselves. They faced the difficult task of making individual decisions about contraceptive use based on statistical probabilities derived from the study of large populations.

The FDA's second report also included new but by no means clarifying information on the relationship between the pill and cancer. Researchers in New York City studied the prevalence of cervical carcinoma in more than 34,000 women attending Planned Parenthood clinics, and found a statistically significant difference between those who chose and used oral contraceptives and those who chose and used diaphragms.[63] The researchers could not ascertain whether the discrepancy resulted from a decreased prevalence among diaphragm users (a possible barrier effect) or an increased prevalence among pill users (a possible hormonal effect). Factors such as age, ethnic origin, socioeconomic status, and number of pregnancies were taken into consideration and found to be inconsequential, but other factors, such as frequency of intercourse, might have been influential in the population with a higher prevalence. The study did not point to a causal role for oral contraceptives in the development of cervical cancer, but only identified a relatively higher prevalence of the disease among pill users than diaphragm users. However, it was easy to misread prevalence for incidence and to assume that the research implied a positive association between oral contraceptives and cervical cancer.

The controversial nature of this paper extended beyond the implications of its ambiguous results. Since the clinical study had been affiliated with Planned Parenthood, that organization's National Medical Advisory Committee requested that the histological slides used as data for the paper be reviewed independently by another group of pathologists, to confirm the diagnoses of the various states of cervical carcinoma.[64] The review revealed discrepancies in both the interpretation of the cytology on the slides and the subsequent classification of the cervical condition as precancerous or not.[65] The editor of the *Journal of the American Medical Association,* to which the authors had submitted the paper for publication, wanted to include the pathologists' report along with the original manuscript and an editorial comment. The senior author, Myron R. Melamed, rejected this proposal, withdrew the paper, and resubmitted it to another journal.[66] It finally appeared in the *British Medical Journal* in July 1969 (one month before the FDA's second report), unaccompanied by the separate review of the pathology. The lay press implied that the editors of the *Journal of the American Medical Association* had suppressed Melamed's paper; one of the independent pathologists argued that the reverse was true, since the pathologists' report disputing Melamed's results never achieved publication.[67]

This large-scale study of the relationship between oral contraceptives and cervical cancer provoked controversy within the medical community both before and after its publication for several reasons. First, the nature and design of the experiment itself were problematic. The focus on prevalence revealed little about the relationship between oral contraceptives and cervical cancer, and by employing diaphragm users as the control group instead of, say, IUD users, the investigators introduced another variable (a possible barrier effect) into the already complicated equation. Second, the interpretation of the data obtained was disputed by other authorities in the field, who differed in their evaluation of the cancerous condition of the cervical cells. Third, by refusing to allow the dissenting report to be published, Melamed and his associates alienated some members of the medical community and caused many others to question their motives and methods. In short, the study added very little to an understanding of the relationship between oral contraceptives and cancer and very much to the flourishing controversy over the health effects of the pill.

About the same time as the Melamed affair, a workshop was held in Boston to address the subject of the metabolic effects of oral contraceptives. Sponsored by the Center for Population Research of the National Institute for Child Health and Human Development (a federal agency) and the Center for Population Studies of the Harvard University School of Public Health, the meeting gathered together fifty-five researchers to present and discuss their work in nine areas: liver and gastrointestinal tract, carbohydrate metabolism, lipid metabolism, protein and amino acid metabolism, respiration, hypertension and electrolytes, calcium and skeleton, vascular system, and central nervous system. The reports were published the following year in a volume of more than 700 pages.[68]

The presentations at the workshop confirmed that anatomical, physiological, and biochemical alterations occurred in virtually every organ of the bodies of women taking oral contraceptives. Individually, the reports of the biological effects of taking the pill presented no immediate cause for alarm because most of the conditions described had few symptoms or none at all (e.g., elevated blood levels of zinc and copper were detectable only by laboratory assays). Taken together, however, they presented disconcerting evidence of a widely used drug with widespread systemic effects, the implications of which were as yet unknown. In the introduction to the published proceedings, the editors noted that "for the most part, the findings are straightforward and noncontroversial . . . the most difficult question to answer relates to the seriousness of such changes."[69] In other words, the facts themselves—biochemical assays, etc.—were not contested by scientists, but their interpretation was subject to debate.

Opponents of the pill appropriated the growing inventory of the metabolic effects of oral contraceptives as evidence that the pill was far too potent an agent to be used for birth control in otherwise healthy young women. However, the review of the extensive medical literature by the task force on biologic effects for the FDA's *Second Report on the Oral Contraceptives*, released in 1969, resulted in a guarded statement of support for the safety of the pill with respect to metabolic effects: "There is no evidence at this time that any of these drug-induced metabolic alterations pose serious hazards to health. The systemic effects of the drugs are so fundamental and widespread, however, that continued medical surveillance and investigation is required."[70]

A spirit of qualified approval permeated the entire second report, which resembled the first report in both format and the makeup of the advisory committee. The most striking difference between the two FDA reports was the explicit notice taken of the public and the press by the committee in 1969. The first report spoke directly to physicians: "In the final analysis, each physician must evaluate the advantages and the risks of this method of contraception . . . He can do this wisely only when there is presented to him dispassionate scientific knowledge of the available data."[71] The second report included the public in its audience: "In the final analysis, both the physician *and the layman* must evaluate the risks of the hormonal contraceptives . . . They can do so wisely only when they have access to all available information, accurately and dispassionately presented. [emphasis added]."[72] The introduction to the chairman's summary contained numerous references to the public and the press: "Since the publication of the last Report on the Oral Contraceptives in 1966, scientific as well as *public* interest in this method of family planning has remained high . . . Adverse reactions are continually reported in the scientific literature and the lay *press* . . . This pharmacological experience is unique also in the attention it has received by the *press* . . . Such reporting is the quickest way to satisfy the *public's* right to know . . . By reviewing a welter of scientific studies of varied value, the *press* has acquired an increasing awareness of the problems . . . So too have physicians and the *public* [emphasis added]."[73] The committee was so concerned with accurate reporting by the press that its list of recommendations included annual conferences of science writers on contraceptive knowledge and advances to be held under the auspices of the FDA.

In the second report, the committee acknowledged the medical controversy surrounding the pill and identified its two main issues: (1) the accumulation of scientific data necessary to establish a cause-and-effect relationship, and (2) the balance between the risks and benefits of taking oral contraceptives. The imperfect voluntary reporting system currently in place hindered data collection, and the risk-benefit calculation involved subjective evaluations. The committee also struggled with the meaning of drug safety, which had been left undefined by the

1962 amendments to the Federal Food, Drug, and Cosmetic Act. The group decided that risk-benefit analysis, despite its subjectivity, had to play a major role in the determination of safety, because no drug could ever be considered 100 percent safe. The final pronouncement on the status of oral contraceptives reflected this limitation: "When these potential hazards and the value of the drugs are balanced, the Committee finds the ratio of benefit to risk sufficiently high to justify the designation safe within the intent of the legislation."

The experts could give the pill only qualified sanction, which was unsatisfying to scientists and physicians, press and public alike. The available data were subject to a wide range of interpretations, and the balance of the known and postulated risks of the contraceptive and perhaps additional health benefits differed for every person. Evidence pointed to a causal role for the pill in the development of thromboembolism, but this correlation remained under contention. The association between oral contraceptives and cancer was still very much up in the air, owing largely to the long latency period for cancer and the relatively recent availability of the pill. Dozens of metabolic reactions to the pill had been documented, but the implications of these effects were not at all clear. By the end of the 1960s, only one thing was certain about oral contraceptives: they were very effective in preventing pregnancy.

Physicians' Responses

The general practitioners and obstetrician-gynecologists who spent their days seeing patients constituted a very different sector of the medical community than those who spent their time doing research and writing journal articles. Practicing physicians may or may not have kept abreast of the latest studies on the health effects of oral contraceptives; their more immediate concerns centered on the needs, demands, and experiences of their patients. Each physician developed his own opinions about the pill, based on the idiosyncrasies of his practice. However, so many articles about the potential health hazards of oral contraceptives flooded both the medical literature and the popular press that it would have been impossible for a doctor not to be exposed to at least some of the controversy.

Some physicians witnessed the adverse effects of oral contraceptives firsthand in their own patients; these doctors tended to be more critical of the pill than their colleagues who had to rely on secondhand reports. All had to weigh these scattered negative incidents against the largely positive experiences of oral contraceptive use among their patients. In the absence of definitive damning evidence against the pill, many doctors were reluctant to stop prescribing it for their patients, especially since this service had proven to be quite lucrative. The enthusiastic reception to the pill among doctors in the early 1960s showed little sign of abatement as the decade progressed.

In 1967, two separate surveys, one undertaken by *McCall's* and the other by *Ladies' Home Journal,* questioned obstetricians and gynecologists about their experiences with and opinions of oral contraceptives. Both polls found highly favorable attitudes toward the pill and widespread prescribing by ob-gyns across the country. The *McCall's* survey, conducted jointly by the American College of Obstetricians and Gynecologists (ACOG), sent out questionnaires to all 8500 ACOG fellows. Of the almost 7000 who responded, 95 percent reported that they regularly prescribed oral contraceptive pills as a method of family planning; less than 1 percent never prescribed the pill. The prescription of oral contraceptives was by no means a unilateral decision by physicians; 96 percent of the doctors said that the pill was the method most frequently requested by their patients.[74] The *Ladies' Home Journal* survey of more than 2500 board-certified obstetrician-gynecologists obtained similar results. Only 1 percent never prescribed oral contraceptives; 89 percent reported that most of their young married patients expected them to prescribe the pills.[75]

The majority of physicians in both surveys felt that women could safely take the pill for four years or longer. Among the doctors in the *McCall's*-ACOG survey, there appeared to be greater concern about the effect of the pill on a preexisting condition than about its causal role in the development of disease. Whereas 83 percent did not prescribe the pill for a patient with breast cancer, only 1 percent attributed the development of breast cancer to use of the pill. Similarly, 76 percent of the doctors considered phlebitis to be a contraindication, but only 21 percent thought that condition would be caused by the pill.[76] Very few of the doctors surveyed by the *Ladies' Home Journal* thought that the pill was too hazardous to use for birth control. While only 3 percent proclaimed the pill to be "safe as water," 52 percent felt that oral contraceptives presented "minimal risks for most women" and another 41 percent felt there were "no harmful effects proved for those who tolerate it."[77] Ninety percent reported that their attitudes toward the pill had remained the same or had become more favorable in the past year. These surveys indicated that in 1967 the pill enjoyed a good reputation and solid backing among those physicians most likely to prescribe it.

However, the increasing reports on the adverse health effects of oral contraception, the lack of consensus on the meaning of these reports, and the concomitant increase in publicity in the late 1960s held some uncomfortable implications for the practicing physician. An editorial in the *Journal of the American Medical Association* reflected on the role of the doctor in the wake of the recently released FDA and WHO reports: "These sharp differences of opinion should be welcomed, since they help to stimulate further research directed toward obtaining answers to these problems. In the meantime, an uneasy responsibility rests with the practicing physician who must properly guide his patients in the use of these preparations without engendering undue alarm, and yet be alert to the possibility that some unusual response may signal a serious adverse reaction."[78]

A year and a half later, another editorial bemoaned the situation of the practicing physician, caught in a web of social, religious, economic, and medical concerns about the drug he was responsible for prescribing to his patients:

> A particularly serious matter to the clinician, the contraceptive pill is unlike any other drug that he prescribes. Not only does it cast him into an unfamiliar role of preventing a natural creative process rather than a disease, but it places him at the mercy of directors who are not always in agreement on how the play is to be enacted. The initiative is obviously not his but the patient's. Nor his to make are the value judgments involved in contraception. Exponents of religious teachings, ethical precepts, sociological principles, and economic law have an important say. Clearly the physician does not hold center stage.
>
> Even when dealing with the pill's medical aspects, the clinician often feels removed from "where the action is." To his patient's simple question, "How safe is the Pill?" he may find no ready answer based on his own experience. So infrequent are medical complications—if they occur at all—that rarely can a practicing physician draw valid conclusions from his own limited observations. Of necessity he must rely on scientists who investigate physiological and biochemical effects of contraceptive substances and on biostatistical experts who analyze data provided by large-scale studies.[79]

This passage identified the main currents in the shifting position of the medical profession with respect to oral contraception. It expressed the continuing ambivalence of the medical profession toward contraception in general. Doctors had always been uneasy about "preventing a natural creative process." Once they decided to co-opt birth control as a medical procedure, they resented the intrusion of lay authorities into their domain. This loss of control was a major concern for practicing physicians. Not only did the patient have a voice in the "treatment" of her condition (namely, the prevention of pregnancy), but also several other "actors" expressed their opinions about the merits and faults of oral contraception. Furthermore, the practicing physician was no longer the sole authority on medical matters. The pill had initially afforded the physician greater medical control of contraception: he had to use his expertise to decide on the best preparation and dosage for each individual patient, who then required continued medical surveillance. Now he was being replaced by research scientists and statisticians whose laboratory data and epidemiological studies provided better answers to questions of the pill's safety than his limited empirical experience.

These wistful editorials did not necessarily influence the practices of the larger medical community. A third survey of obstetrician-gynecologists, conducted in 1970, found that the overwhelming majority of physicians continued to prescribe oral contraceptives in spite of all the negative publicity.[80] The results of this poll indicated that physicians did not react strongly to the media blitz about the pill; the few who opposed its use had done so for many years and the large majority who condoned its use had not been swayed by the recent controversy.

Indeed, despite the incursions into his territory, the doctor still exerted a fair degree of medical control in the realm of family planning; women had to obtain the physician's prescription, and thereby his permission, in order to use birth control pills. The pill was still the easiest and most lucrative method of contraception for a physician to recommend; writing prescriptions to be filled by the pharmacist took far less time and effort than fitting and inserting IUDs and diaphragms. Given the wide divergence in opinion among medical researchers as to the safety of the pill, the practicing physician may have felt perfectly justified in continuing to prescribe oral contraceptives for his patients, most of whom were grateful and willing users, until incontrovertible evidence appeared.

Women's Responses

Of course, not all physicians were enthusiastic advocates of oral contraceptives; likewise, not all women were eager to use them and, of those who did take the pill, not all were satisfied with their experiences. Those who suffered serious side effects often blamed their physicians for negligent treatment. The *Ladies' Home Journal* was flooded with responses to its 1967 article, "The Terrible Trouble with the Birth-Control Pills," many of which detailed personal ordeals:

> I took the pills for a year and a half. In this year and a half I switched to three different kinds because I got sick. All this time I was under a doctor's supervision. I would call my doctor and tell him I thought the pills were making me sick, so he would switch me to another pill.[81]

> I was sitting in the waiting room of the intensive-care unit . . . when I read your article on the dangers of the pill. It was a little late for me to read it since my wife had just been operated on for massive clotting in her leg. She had been taking the pill for five months when the attack occurred. She had been ill for two weeks with sharp pains traveling through her body, and we had been to our family doctor twice and he had not found the cause of her illness.[82]

For these people, this magazine article reported what they already knew and experienced firsthand about the pill's health effects. For others, media coverage introduced them to the controversy over the safety of the pill and influenced their opinions: "For what it's worth—if you keep track of such things—I'm off the pill because of your article in the July issue. The convenience just wasn't worth the risk."[83]

Planned Parenthood officials worked hard to counter the negative publicity about the health effects of the pill, which they believed was a tactic employed by "competing publications as part of the 'rat race' for circulation and survival in a highly competitive market."[84] The medical directors released a three-page statement in rebuttal to the "Terrible Trouble" article in which they declared: "A recent article in a leading women's magazine has compiled negative data out of context and has presented a one-sided view that is misleading and needlessly alarming."[85] They went on to reassure women that the risk of death from taking the pill was statistically very low and that the pill continued to be the most effective method of preventing pregnancy. Five months later, in a follow-up response to "the unfortunate scare article," the Medical Committee "help[ed] create and [took] advantage of the opportunity to infuse some radio and television programs with factual and mainly affirmative background on the pill."[86] When the media coverage of the pill controversy showed no signs of abating, Planned Parenthood considered the possibility of establishing an annual award—with a prize of $10,000—to honor journalists who demonstrated "excellence in communications about contraception."[87]

Planned Parenthood's fear that women would be scared off the pill by sensationalist articles was not unfounded. Demographers found that the proportion of married women using the birth control pill leveled off at about 30 percent and the dropout rate increased in the late 1960s. Closer analysis revealed that 70 percent of oral contraceptive dropouts quit the pill as a result of physical problems experienced personally and 17 percent quit because they were worried about the reports of adverse health effects. The authors of the 1970 National Fertility Study concluded: "It seems plausible that the sustained rise in the dropout rate that occurred in the late 1960's despite improvements in the product which reduced side effects was due to the anxiety about serious health implications generated by responsible medical reports and sometimes exaggerated publicity."[88] Two Gallup polls, taken three years apart, recorded a definite shift in public opinion. In December 1966, in response to the question "Do you think these pills can be used safely—that is, without danger to a person's health?" 43 percent said yes, 26 percent said no, and 31 percent had no opinion. When asked the same question in February 1970, only 22 percent thought the pill was safe; 46 percent thought it was unsafe, and 32 percent were unsure.[89] Respondents had taken to heart the message presented by three years of critical reporting on the questionable safety of oral contraceptives.

Although the rate of increase in the number of new prescriptions slowed down, millions of American women continued to use oral contraceptives. Many felt that the advantages conferred by the effectiveness of the pills in preventing pregnancy outweighed any disturbance caused by side effects or the risk of more serious medical consequences. One woman explained her personal dilemma: "I feel as if I'm going against nature. I know these pills are fighting other elements in my body . . . [but] It's the safest method of contraception. The only one I trust almost 100 per cent."[90] Other women wanted to learn more about the potential consequences of taking the pill and sought information from their physicians. Unfortunately, given the unresolved status of the medical debate over the safety of the oral contraceptives, doctors did not have answers to their patients' questions.

Doctors did, however, have opinions, and it was perhaps the expression of those diverse opinions that contributed to the deterioration of the relationship between doctors and patients. Faced with myriad potential health effects of oral contraceptives, some physicians discouraged their patients from taking the pill. In her book, *The Doctors' Case against the Pill*, Barbara Seaman described dozens of doctors who questioned the safety and merit of oral contraception. Certainly some women respected their doctors' judgment and heeded their advice. For others, however, the choice was not so simple. Women who had been unsuccessful with other methods of birth control were reluctant to give up the pill and the freedom from pregnancy that it conferred. Seaman quoted one satisfied pill user's response to her doctor's warnings: "I don't care if you promise me cancer in five years, I'm staying on the pill. At least I'll enjoy the five years I have left. For the first time in eighteen years of married life I can put my feet up for an hour and read a magazine . . . If you refuse to give me the pill, I'll go get it from someone else."[91]

Other doctors tailored their advice to patients based on the woman's socioeconomic status. They figured that middle-class women were conscientious enough to be able to use "safer," albeit less effective, barrier methods of birth control, but that poor women needed both the effectiveness and the simplicity (i.e., coital independence) of the oral pill.[92] For these doctors, medical decisions about contraception blurred into sociological decisions about family planning and social control. Although women could read the same articles in magazines and newspapers on the medical controversy over oral contraceptives, they received very different medical counsel based on the physician's assessment of their income and educational status.

Still other physicians regarded the controversy over the safety of the pill as blown way out of proportion, mainly because of sensationalist media coverage, and resented their patients' entreaties for better information on the risks involved in taking the pill. One obstetrician who participated in the *McCall's*-ACOG survey remarked: "Some women just aren't comfortable on oral contraceptives. . . . These are the women who make the phone calls that drive us crazy."[93] This physician, in dismissing patients' concerns as silly, was less inclined to share with them the conflicting medical evidence on oral contraceptives, and thus effectively reduced their ability to participate in making informed decisions about contraceptive choice. Furthermore, women who expressed their concerns about side effects to such physicians ran the risk of being labeled as "difficult" patients.

The importance of the media in publicizing the controversy over the safety of the pill cannot be overestimated. The space given to oral contraceptives as the subject of both headline news and feature stories focused the public's attention on the problem. Most important, by narrating women's stories about their experiences, mass media articles reduced the isolation of individuals' ordeals and created a sense of shared experience among pill users. Women who had suffered physical or psychological reactions while taking the pill could identify with the women described in the magazines. They realized that their troubles were not unique, and that the pill might have been responsible for their ailments. Women who had not had any problems with the pill learned about the negative experiences of others and began to question the safety of the drug and the wisdom of using it. As journalists reported medical researchers' quest for more information to assess the relative risks and benefits of oral contraception, women turned to their links to the medical world—their physicians—for more information to make personal decisions about the pill.

The gap in the medical profession's understanding of all the health implications of oral contraception in combination with the media blitz on the subject created tension over how much and what kind of information ought to be imparted in the doctor's office. The problem cannot be simply reduced to patients demanding information that physicians were withholding, although that scenario may be accurate for some doctor-patient interactions in the late 1960s. In other cases, doctors tried to tell patients things that the patients didn't want to hear, or patients didn't think to ask questions, or they did ask but the physicians didn't know the answers. When women compared notes on their experiences with doctors and contraceptive advice, they discovered a variety of opinions, recommendations, and practices. The critique of the male medical profession and its treatment of women had not yet been articulated by the new generation of feminists, but the seeds of this analysis germinated in the frustrating interactions between women and their doctors in the absence of definitive data on the health effects of the pill.

Chapter Five

Oral Contraceptives and Informed Consent

In September 1969, the head of the information and education department of Planned Parenthood sent memos to the directors of the medical department and to the president of Planned Parenthood about the imminent publication of a book called *The Doctors' Case against the Pill* by Barbara Seaman. The memos warned that "the effect of the book . . . will be to destroy consumer faith in the efficacy of the pill" and advised the Planned Parenthood leadership to "put balance and background in the hands of reviewers to offset the impact of this book in all media before the bomb hits."[1] The following month, days before the scheduled release of Seaman's book, Planned Parenthood sent a memo to all of its affiliates' executive directors and medical directors to alert them to "books attacking the pill." The memo contrasted portions of text from *The Doctors' Case against the Pill* with sections of the 1969 FDA Advisory Committee's *Second Report on the Oral Contraceptives* to illustrate the "distorted picture" presented by Seaman. It urged the affiliates to respond to queries from local media by "report[ing] your own clinic[']s experience to add the very impressive positive side of oral contraception."[2] What did this book say to elicit such a strong reaction from the nation's largest family planning organization?

The Doctors' Case against the Pill presented a wealth of evidence against the safety of oral contraceptives. Barbara Seaman had written dozens of articles during the 1960s about issues related to women's health and well-being for magazines such as *Bride's* and *Ladies' Home Journal*. Well versed in the journalist's tactics to grab and hold the reader's attention, Seaman marshalled the testimony of physicians, medical researchers, and women who had used oral contraceptives to build her case against the pill and to indict the medical-pharmaceutical establishment that developed and marketed it.

Seaman's critique of hormonal birth control in particular and the medical–pharmaceutical complex in general lent support to the two renascent social movements of consumerism and feminism. The consumer movement, which originated in the United States with efforts to regulate the manufacture of food and drugs in the late nineteenth century, entered a new phase in the 1960s with the publication of books such as Rachel Carson's *Silent Spring* (1962), Jessica Mitford's *The American Way of Death* (1963), and Ralph Nader's *Unsafe at Any Speed* (1965).[3] Whereas Carson took on the chemical industry, Mitford the funeral industry, and Nader the automobile industry, Seaman challenged the pharmaceutical industry and the closely allied medical profession. She demanded that pill manufacturers and physicians share all available information about the health risks of oral contraception with patients, so that the women themselves could make informed decisions about birth control. This demand for full disclosure represented an extension of the relatively new concept of informed consent in medicine beyond the operating room to include all doctor–patient interactions.[4]

Seaman's charge that women received inadequate care from their physicians, particularly obstetrician-gynecologists, mirrored a growing concern among feminists about women's health issues. As women sought to gain control over various aspects of their lives, they realized that they first needed to gain control over their own bodies. While Seaman was writing her book in New York, the women who would soon form the Boston Women's Health Book Collective were meeting in Massachusetts. *The Doctor's Case against the Pill* inspired feminists to vocalize the shared perception that the medical profession was "condescending, paternalistic, judgmental and noninformative."[5] In part owing to the publication of this book, the controversy over the safety of the pill galvanized the women's health movement of the 1970s.

Feminist-activists first became involved in the pill debate in response to hearings held in January 1970 by U.S. Senator Gaylord Nelson's Subcommittee on Monopoly of the Select Committee on Small Business. Although the feminists protested at the hearings, in fact they shared with Senator Nelson the same concern that the pharmaceutical industry and the medical profession were being less than honest with women about the hazards of oral contraceptives. Nelson's hearings marked the entry of the federal government into the medical and public controversy over the safety of the pill.[6] In less than a year, that intervention turned into regulation, as the Food and Drug Administration ordered the manufacturers to include inserts for patients describing the known health risks of oral contraception in every package of birth control pills.

The battles over the inclusion of the patient package insert and its wording identified differing intentions among groups purportedly working toward the common goal of better family planning. Pharmaceutical companies, physicians, population control and family planning organizations, and feminist groups expressed their interests as the federal government planned to extend its regulatory authority to include oral contraceptives. Close examination of the Senate hearings on the pill and the subsequent development of the package insert reveals the conflicting missions of the various groups with interests at stake in the debate over the safety of the pill.

Hearings on "Competitive Problems in the Drug Industry"

In 1994–95, Barbara Seaman celebrated the twenty-fifth anniversary of *The Doctors' Case against the Pill*. Publications ranging from *Science* to the *Boston Phoenix* marked the occasion with articles on birth control and women's health issues.[7] A noted female artist hosted a party at her home in fashionable East Hampton, New York, to honor the author, and Representative Jerrold Nadler of New York "salute[d] Barbara Seaman as a national role model" on the floor of the U.S. Congress.[8]

The book created less of a stir when it was first released in 1969. Seaman recalls that "it didn't make a big splash when it first came out, not in the first few weeks."[9] When the *New York Times* finally got around to reviewing the book, two months after its release, the reviewer dismissed the book as "disorganized" and "scatterbrained."[10] However, the publisher, Peter Wyden, believed that the message of the book deserved a wider audience. He hired a publicist in Washington who passed along a copy of *The Doctors' Case against the Pill* to Senator Nelson.[11]

By the time Nelson received Seaman's book, he had already led his subcommittee through sixty-four days of hearings over two and a half years. The hearings filled fourteen volumes with testimony, inquiry, and supplementary materials on a broad range of topics, including drug pricing, testing, and advertising; brand name versus generic drugs; the relationship between the drug industry and the medical profession; and investigations into the use and misuse of specific kinds of drugs, such as psychotropics (barbiturates and tranquilizers) and combination antibiotics.[12] At the start of the hearings, Nelson announced that "these hearings are concerned with the important matter of the health and pocketbook of American citizens."[13] One year earlier, the U.S. Congress had approved the

Medicaid and Medicare amendments to the Social Security Act, which used federal tax dollars to provide health care for the poor and the elderly, respectively. Since the federal government spent more than half a billion dollars annually to purchase prescription drugs, Nelson, as a member of the legislative branch, felt it was his duty to expose the pharmaceutical industry to public scrutiny.[14]

After the first year of hearings, Nelson concluded that the current state of affairs in the drug industry presented two major problems to the health and well-being of the American people. First, both the medical profession and the Food and Drug Administration depended on pharmaceutical companies for information about drugs. Since the goal of the industry was to sell drugs and to make money, neither doctors nor government officials could rely on promotional literature to be completely objective; each company tried to portray its products in the best possible light. Nelson charged that the FDA was "totally dependent—at present—on the industry which it was established to regulate," and therefore could not provide doctors with unbiased information about prescription drugs.[15] The second problem arose because the industry, not the FDA, was responsible for conducting and/or arranging for tests and clinical trials of new drugs. Since pharmaceutical firms wanted to earn FDA approval quickly for their products so that they could reach the market as soon as possible, they often compromised the quality of new drug evaluations.

Six months later, Nelson added a third problem to his growing list of concerns. Not only did the pharmaceutical industry wield too much power in its relationship with its regulatory agency, but it also exerted significant control over the medical profession. Nelson questioned the professional and ethical implications of close relations between medicine and the drug industry as in, for example, industry-sponsored medical publications and physician-stockholders in drug companies. Since doctors relied in part on drug detailmen, promotional literature, and free samples provided by drug companies for their continuing education in pharmacology, Nelson worried that physicians could not make intelligent decisions about prescription drugs because of the commercially biased information they received.

Thus when a book describing the hazards of a drug taken by millions of women each day landed on his desk, the senator sat up and took notice. He and the staff economist for the Senate Committee on Small Business, Ben Gordon, interviewed Barbara Seaman several times during the fall of 1969. Seaman recalled: "I spent a lot of November and December going back and forth to Washington, being sort of grilled. They wanted to make sure that I wasn't a nut. They wanted to make sure that I wasn't too frivolous—a women's magazine writer or somebody just trying to sensationalize it [the pill]. Maybe they also wanted to make sure that I wasn't a combat booted feminist."[16] Seaman passed the test, and on December 22, 1969, Nelson released a statement to the press announcing his intention to hold public hearings on the oral contraceptives. He restated his mission the following month on the first day of the hearings: "The aims of these hearings . . . are to present for the general public's benefit the best and most objective information available about these drugs. First, whether they are dangerous for the human body, and, second, whether patients taking them have sufficient information about possible dangers in order to make an intelligent judgment whether they wish to assume the risks."[17]

The Pill Hearings, Part I

The two issues of oral contraceptive safety and informed consent corresponded closely to Senator Nelson's larger concerns about medicine and the pharmaceutical industry. The adverse health effects of oral contraceptives cast doubts on the merit of the drug tests and clinical trials conducted prior to FDA approval, and thus deepened his suspicions about the integrity of the drug manufacturers. The question of whether women could make informed decisions about using the birth control pill reflected Senator Nelson's misgivings about the ability of physicians to obtain objective information from the pill manufacturers and in turn to pass that information on to their patients.

To evaluate the alleged health risks of oral contraceptives, Nelson and Gordon assembled a group of expert witnesses to testify about the biochemical, physiological, and psychological effects of the pill. Many of the scientists and physicians who appeared before the committee had also appeared on the pages of Seaman's book, and their testimony provided little, if any, new information about the biological effects of oral contraceptives. However, as a result of the intense media coverage of the hearings, the medical controversy over the safety of the pill reached a much wider audience. As Nelson commented at the end of the hearings, "although very little of the information presented here or perhaps none of it was new to the experts in the field, quite obviously a lot of it was not known to the practicing physician who prescribes the pill and the public which consumes it."[18]

Although Gordon relied heavily on Barbara Seaman for advice in choosing witnesses for the hearings, he did not invite her to appear before the committee, "because she wasn't a primary source."[19] Nor were any women who had experienced pill-related adverse health effects asked to testify, because "Nelson . . . didn't like that way of doing things . . . He wanted to keep the hearings on a high level."[20] However, in earlier hearings on the fatal side effects of the antibiotic

Chloromycetin, the committee heard the testimony of three fathers (two physicians, one newspaper publisher) whose children had died as a result of taking the drug, and therefore a precedent did exist for the inclusion of lay witnesses. During the pill hearings, feminists denounced Nelson for allowing only the testimony of physicians and scientists, most of whom were male.

Seaman, by contrast, recognized the importance of women's voices and experiences in the pill debate, and she included in her book heart-rending stories of women who had died or suffered serious consequences as a result of taking oral contraceptives, along with the expert testimony of scientists and physicians. She had hoped to call the book *The Case against the Pill,* but at the last minute found out that another science writer intended to use that title (his book never appeared in print). Reluctantly, Seaman agreed to capitalize on the respectability of her experts and named the book *The Doctors' Case against the Pill.*

In a way, Seaman and Nelson represented the ambivalence toward science and medicine in American society at the end of the 1960s. On the one hand, they questioned the merit and safety of the pill, a product of medical science and technology, and criticized scientists and physicians for their incursion into family planning. On the other hand, they relied on scientists and physicians for evidence on which to base their critique. Nelson never acknowledged this inconsistency, at least not in public. Seaman, however, underwent a significant transformation after the publication of *The Doctors' Case against the Pill.* She had written the book as a "reformist" feminist, hoping to influence changes within the existing system of medical care. Later, after coming in contact with more radical feminists, she became critical of the medical establishment and its (mis)treatment of women's health care needs.

Seaman's enlightenment began on the first day of the Nelson hearings. Although not invited to testify, she attended the hearings as a press correspondent, on assignment from the North American Newspaper Alliance. In the middle of testimony from the second witness, some women in the audience interrupted the hearings. Seaman vividly recalled the disturbance:

> All of a sudden, these women started standing up and yelling . . . I heard my name, "why isn't Barbara Seaman testifying?" And then somebody else was saying, "why isn't there a pill for men?" And someone else was saying, "why aren't there any patients testifying?" . . . Then they were cleared out of the room . . . So, I went outside. I followed them out. This was my story. What these guys were saying that I'd been writing about for years wasn't my story . . . So I went out and introduced myself to them . . . When I said that I was Barbara Seaman, they fell all over themselves . . . It turned out that it was my book that had inspired them to demonstrate.[21]

The demonstrators belonged to a group called D.C. Women's Liberation. The group initially came together in 1969 to protest illegal abortion; their work soon led them to a broader interest in women's health care in general. Alice Wolfson, a member of D.C. Women's Liberation and later one of the founders of the National Women's Health Network along with Barbara Seaman and two other women, explained how the demonstration got started:

> What happened was that we were sitting at a meeting—which we did all the time in those days, nobody worked, I don't know what we did for money!—in the middle of the afternoon, and we had heard that there were going to be these hearings on the birth control pill on the Hill, so about seven of us went . . . When we got there, we were both frightened, really frightened, by the content and appalled by the fact that all of the senators were men [and] all of the people testifying were men. They did not have a single woman who had taken the pill and no women scientists. We were hearing the most cut-and-dried scientific evidence about the dangers of the pill . . . Remember all of us had taken the pill, so we were there as activists but also as concerned women . . . So while we were hearing this, we suddenly said, "my God," and we—because in those days you did things like that—raised our hands and asked questions.[22]

After this spontaneous outburst, the women quickly organized themselves and held militant demonstrations at every subsequent hearing on the pill. A group of twenty women arrived early each day and strategically placed themselves at the end and in the middle of every other row of seats. They came prepared with questions to interrupt the hearings and with bail money tucked inside their boots.

Ironically, both the feminists and Senator Nelson agreed on most issues concerning the pill. They believed that the FDA had allowed the drug companies to market the pill without adequate tests of its long-term safety. To illustrate the problem of insufficient testing, both likened the millions of women who used the pill to unsuspecting guinea pigs in a massive experiment. In spite of the medical controversy over the safety of the pill, neither the feminists nor Nelson advocated a ban on the oral contraceptives; both realized that such a proposal was unrealistic and impractical, and could lead to a black market for the pills that would be even more dangerous to women. Instead, both Nelson and the D.C. Women's Liberation group argued that women needed access to all available information on the pill so that they could make intelligent decisions about birth control. They agreed that the lack of informed consent stemmed from the problem of poor communication between doctors and patients. At this point, the opinions of the senator and the feminists diverged. For Nelson, the issue of informed consent could be solved by improving the channels of communication among the manufacturers, the FDA, physicians, and patients. The more radical feminists saw the pill as the tip of the iceberg of much larger problems in women's health care; they doubted that these problems could be solved within the context of the contemporary system of male-dominated medicine.

Viewing the hearings through this particular lens, the members of D.C.

Women's Liberation took umbrage at the absence of female witnesses. Nelson's attempt to placate the demonstrators after the first interruption—in which he referred to them as "girls" and counseled them against disruptive behavior—only exacerbated the women's anger and increased their resolve to protest. Although they agreed with Nelson on the importance of increasing public awareness of the health risks of oral contraceptives, they could not tolerate the established format of the Senate hearings.

Nelson also faced criticism from within his Senate subcommittee, mainly from a freshman senator from Kansas named Bob Dole. Dole professed concern that women would be unduly alarmed by public hearings of medical testimony regarding the safety of the pill. In his opening statement (read in his absence by the minority counsel), Dole warned: "We must not frighten millions of women into disregarding the considered judgments of their physicians about the use of oral contraceptives . . . Let us show some sympathy for the beleaguered physician who must weigh not only the efficacy and safety of alternative methods for a particular woman, but the emotional reactions of that woman which have been generated by sensational publicity and rumored medical advi[c]e."[23] As a conservative Republican, Dole's ulterior interests probably agreed with those of the pharmaceutical industry, as well as the "beleaguered physician." Later, when he joined the hearings on the third day, he remarked: "There must be certain advantages to the pill other than avoiding pregnancy. I think we probably have terrified a number of women around the country . . . I would guess they may be taking two pills now—first a tranquilizer and then the regular pill—because of our erudite investigation."[24]

The controversial nature of the pill hearings did not go unnoticed by the media. All three national television networks covered at least four of the five days of the first set of hearings on their evening news broadcasts in January 1970; NBC offered additional coverage and commentary on several days when the hearings were in recess. Of the eight witnesses who testified during the first two days, seven voiced serious concerns about the safety of oral contraceptives, and excerpts from their testimony were televised to the American public. Those who tuned in to ABC heard Sam Donaldson report, "Doctors do not agree on the relative safety of the pill. But on this first day of Senator Gaylord Nelson's hearings, the emphasis was on the dangers. Dr. Hugh Davis is sufficiently alarmed to suggest that no woman use the pill for more than two years." Viewers then heard a portion of Hugh Davis's testimony, in which he said, "These agents are somewhat like an iceberg. The obvious problems have surfaced in the form of blood clotting disorders. A nagging specter of cancer remains."[25] On CBS, Walter Cronkite reported, "Almost nine million women in America, and ten million elsewhere, are taking the pill each day, in the words of one expert, 'as automatically as chickens eating corn.'"[26]

The Washington reporter David Dick summarized a different portion of Hugh Davis's testimony: "Dr. Hugh Davis, of Johns Hopkins University, testified that the possible side effects are so great, if the pill were a food product it would probably be ordered off the market."[27] NBC's Chet Huntley offered a third view of Davis's testimony in his introduction to the lead story of that network's evening news: "Dr. Hugh Davis, of Johns Hopkins, told the committee, 'Never in history have so many individuals taken such potent drugs with so little information as to actual and potential hazards.'" Huntley added, "The pills have been on the market for ten years, but today was the first time Congress has seen fit to investigate them."[28] The television cameras also captured the disturbance created by the women from D.C. Women's Liberation as well as the impromptu news conference held by the women after the hearings. A Women's Liberation spokeswoman asserted, "All of the women here have suffered ill effects of the pill. And we were told by the doctors, while suffering these effects and afterwards, to go on taking the pill."[29]

The televised newscasts of day one of the hearings provided the public with sound bites of official testimony from so-called medical experts, unofficial testimony from women who had experienced side effects from oral contraceptives, and editorial comment on the controversy over the pill. Anyone who had not previously been aware of this growing controversy was immediately clued in by watching the evening news on January 14, 1970. The first minutes of televised news about the pill hearings portrayed several of the conflicts surrounding the oral contraceptives. First, the debate within the medical profession over the safety of the pill was highlighted, with emphasis on the critics of oral contraception. Second, the confrontation between D.C. Women's Liberation and the all-male Senate subcommittee was presented, demonstrating the tensions between the budding feminist movement and the existing establishment in government, industry, and medicine. Third, the statements of the members of the women's liberation movement revealed the lack of communication between doctors and their female patients.

In the following days of hearings and televised news coverage, the divergent positions of groups with a stake in the fate of oral contraception became apparent. Despite repeated invitations from Nelson, no industry representative ever appeared before the committee. G. D. Searle & Company chose to defend the pill outside the Senate Caucus Room by allowing one of its spokesmen, Dr. Irwin Winter, to be interviewed on television in his office at Searle's headquarters in Skokie, Illinois. Dr. Winter complained, "I thought when we brought out the pill several years ago that we were doing mankind [sic] a great favor and that this is a thing that women had been wanting from time immemorial. Now, from the headlines and the way this is being reported, or being presented . . . one would think we had unleashed a monster on the world."[30]

Nelson's invitations to the drug companies contrasted dramatically with his disregard of the feminists who had asked to testify. On January 23, an organized faction from D.C. Women's Liberation again interrupted the hearings, demanding to be allowed to appear before the committee. All three networks featured the protest on the evening news, in which different women in the audience shouted:

> Why have you assured the drug companies that they could testify? Why have you told them they will get top priority? They're not taking the pills, we are!

> Why is it that scientists and drug companies are perfectly willing to use women as guinea pigs in experiments to test the high estrogen/low estrogen content of the pill, but as soon as a woman gets pregnant in one of these experiments she's treated like a common criminal! She can't get an abortion!

> Women are not going to stay quiet any longer! You are murdering us for your profit and convenience![31]

ABC reported that after Senator Nelson ordered the room to be cleared, he readmitted the press, but "the public and its virago element were not welcomed back."[32] The other networks were more generous in their characterizations of the demonstrators and their intentions.

The uncomfortable role of the FDA as mediator between the drug industry and the public became apparent during the course of the hearings, as several of the witnesses referred to the 1969 *Second Report on the Oral Contraceptives* by the FDA's Advisory Committee on Obstetrics and Gynecology, which declared the pill to be safe. NBC presented part of the testimony of Dr. Victor Wynn, who pointed out the worldwide significance of the FDA's assertion of the pill's safety: "The women in this country, and the women in Great Britain, and in Scandinavia, and in Australia, and in South America, are consuming oral contraceptive medications in the belief that they are considered safe by the American Food and Drug Administration and for no other reason, because they have no other way to assess the safety of this medication."[33]

During this first round of hearings, in interviews with news correspondents, the commissioner of the FDA, Dr. Charles Edwards, stood by the advisory committee's positive assessment of the pill's safety and offered the following counsel to women on the pill: "she should continue to take it . . . through her physician and through the press, she should keep informed as to the hazards, the risks in taking the pill . . . I think if she will do that, then I think without any question that at this point in time we could not recommend that all women stop the pill."[34] To facilitate physicians' knowledge and communication of the adverse health effects of oral contraceptives, the FDA sent out letters to 381,000 doctors, hospitals, and pharmacies.[35] The letter encouraged doctors to tell women about the risks of taking oral contraceptives, especially the recent finding about the link between the pill and thromboembolism. Later, Dr. Edwards and others at the FDA would become skeptical of the ability of physicians to provide sufficient information about oral contraceptives to patients, and would suggest the requirement of an information pamphlet for patients in every package of birth control pills.

In spite of their central importance in the question of the safety of oral contraceptives, women—with the exception of the protesters—did not actively participate in the hearings. After the second day of testimony from physicians and medical researchers, one network thought to seek out women of reproductive age, who had to confront the issues of birth control each day. When an NBC correspondent interviewed several women at the Planned Parenthood clinic in Chicago, he found that "there were various reactions among women about the pill, but no clear-cut consensus."[36] All of those interviewed framed the question of contraception in terms of relative risk; each individual weighed the risk of pregnancy and the risk of adverse health effects associated with the pill differently. It remained to be seen how the millions of pill users would respond to the information spread by the Senate hearings and the attendant news coverage.

The national leadership of Planned Parenthood moved quickly in an attempt to allay the fears of clinic patients and the general public in the wake of the first round of hearings. Even before the hearings began, the organization had mobilized its public relations forces to counteract what they correctly anticipated to be harsh criticism of the pill and the physicians who prescribed it. One memo warned that "the hearings will undoubtedly alarm women across the nation now on the pill—and may even result in its being removed from the market."[37] Although fewer than 5 percent of the American women taking oral contraceptives received their prescriptions at a Planned Parenthood clinic, more than 75 percent of Planned Parenthood patients chose the pill for birth control. The organization's officials feared that women who abandoned the pill would not switch to another method, which would result in an increase in unwanted pregnancies.

The objectives of the publicity campaign were twofold: to continue to meet the birth control needs of patients and to preserve the reputation of Planned Parenthood and its physicians. To achieve these goals, the organization undertook: "a three-pronged approach: We try to educate the media, both national and local; we try to prepare physicians to speak up with one positive voice; we try to keep our Affiliates constantly informed so they can respond responsibly."[38] The medical department, the information and education department, and the president, Dr. Alan Guttmacher, sent letters to Planned Parenthood's affiliate clinics before, during, and after the hearings. These notices reminded the clinics' directors of the organization's long-standing policies regarding the birth control pill. On January 8, a memo emphasized three points: "1. Our clinics offer patients a choice from the *full range* of contraceptive methods. We are not "pill clinics". 2. Our patients are medically supervised . . . No long-term prescriptions are written. This makes it obligatory for women using the pill to receive renewals from a physician. Patients are instructed to query the physician in the event they experience unexpected side effects. 3. Patients are advised of the advantages and disadvantages of all methods, the benefits and the risks."[39] Two days after the start of the hearings, Alan Guttmacher sent a memo to reassure the affiliates that no new medical evidence had been presented, to inform them that Planned Parenthood's position on the pill remained unchanged, and to invite them to call or write to him with questions or concerns.[40] Two weeks later, the medical department reiterated Guttmacher's reassurances in yet another memo.[41]

Guttmacher and his staff made it very clear to anyone who would listen that Planned Parenthood maintained a strict policy of informed consent for all its patients. They also took pains to emphasize their consideration for the individual woman. One administrator wrote to another: "Naturally we are interested in the population problem but I think we should very forcefully declare that we consider the pill safe with regard to its use by the average American woman, not merely because it appears to be one of the most effective methods of mass programs."[42] The family planning advocates most certainly did not want to be represented as callous population controllers, particularly when the majority of their clientele came from families of low socioeconomic status. At the same time, Planned Parenthood was committed to the prevention of "social pathology associated with unwanted conception."[43]

Nelson invited Guttmacher to testify before the committee during the second round of hearings in February 1970. In an article for the *Wellesley Alumnae Magazine,* Guttmacher complained, "By that time news media will be so jaded by the pill story that those of us who are favorable to its use will get little or no coverage."[44] Nevertheless, he and his staff carefully prepared his statement to counterbalance the "barrage of anti-pill publicity" presented during the earlier hearings.[45] One senior staffperson commented that Nelson "feels that he and Barbara Seaman are Rachel Carson and that, in a few years, we will have the phenomenon of the pill being in the same position as DDT," and suggested, "Our stance at this time might be that Senator Nelson obviously had every intention to perform a useful public service, but the realities are that, for whatever reasons, the hearings have spread panic and confusion among women."[46] Guttmacher sought to reassure those panicked and confused women. By minimizing the health risks of oral contraception relative to its benefits in preventing pregnancy, Guttmacher appeared both in print and in person as a strong advocate for continued use of the pill.

Guttmacher's pro-pill stance placed him on the opposite side of the battle line drawn by D.C. Women's Liberation over the pill.[47] However, the Planned Parenthood leader was actually a great ally of health feminists in the related cause of abortion rights. In his testimony to the Nelson subcommittee, Guttmacher argued that the pill was a major weapon in the struggle to reduce the number of deaths associated with illegal abortions.[48] Soon after the pill hearings, he became an outspoken advocate for the legalization of abortion.[49] In early 1970, however, his attention was focused mainly on the matter of the oral contraceptives.

In spite of the efforts by Guttmacher and Planned Parenthood to reassure women that it was safe to continue using the pill, many remained unconvinced. By the end of the first round of hearings, the television-viewing public knew a great deal more about the controversy surrounding the safety of the pill, but no more about whether the pill was safe to take. The *Washington Daily News* captured this uncertainty in a cartoon of a woman in a Hamlet pose holding in one hand a birth control pill and in the other a newspaper headlined, "Scientists say pill is unsafe but safe enough." The cartoon's caption read: "To take or not to take . . . "[50] On February 9, *Newsweek* reported the results of a Gallup poll of women between the ages of twenty-one and forty-five.[51] News of the hearings reached an extremely wide audience; 87 percent of American women had heard or read about them. The survey found that 18 percent of the eight and a half million women with pill prescriptions had stopped taking the pills in recent months, and another 23 percent were considering stopping. One-third of those who had quit or thought about quitting directly attributed their recent or imminent abandonment of oral contraceptives to the Nelson hearings; another one-fourth cited side effects—experienced personally or by friends—as the reason for their doubts about the pill.

Perhaps the most disturbing finding of the survey addressed one of Senator Nelson's initial questions in the pill hearings: Were women being adequately informed by their doctors about the adverse health effects of the pill? The answer was a resounding "no." The poll revealed that two-thirds of women on the pill were never told by their physicians about the potential health risks of oral contraception. Millions of women chose to take birth control pills without knowing the whole story; the lack of communication between doctor and patient precluded informed consent in decision-making about birth control. This discrepancy between their doctors' actions and the expectations of the Senate committee heightened women's concerns about the wisdom of taking birth control pills in particular and about the quality of their medical care in general.

Data from Planned Parenthood clinics mirrored the trend away from the pill identified by the *Newsweek*-Gallup poll. In Detroit, during the month from January 14 to February 13, the clinic reported a tenfold increase in diaphragm requests, a doubling of requests for IUDs, and an "astounding increase" in requests for tubal ligations and vasectomies, all resulting from patients discontinuing the pill. At the New York City clinic, almost one-fifth of oral contraceptive users switched to either the IUD or the diaphragm.[52] *Business Week* reported an "almost hysterical cry for diaphragms" in January and February, resulting in a fivefold increase in sales of the device.[53] A urologist in Rhode Island reported that vasectomy requests had tripled since the hearings began.[54] Women all over the country telephoned their physicians demanding information on whether they should continue to take the pill.

Some doctors, including Alan Guttmacher, regretted the public forum of the hearings because of the alarm generated. An article in the *New York Times,* written by a physician, described the results of the hearings as "an epidemic of anxiety that has spread like wildfire." The author likened the scientific debate in the Senate Caucus Room to medical rounds in a hospital: "Unless the physician explains to the patient in words of one syllable and with understanding and sensitivity the facts in simple lay language, the consequences are anxiety and often real fear . . . The point is that when scientific debate occurs in the public forum and the debate is before a forum of nonscientists, interpretation by the public is often one of general confusion."[55] He assumed that the average person could not understand medicoscientific discourse, and therefore should not be privy to such discussions. This attitude characterized the mindset of traditional medicine, in which the wise doctor made all of the decisions for the compliant patient. Since the introduction of the pill ten years earlier, physicians had arrogated birth control to themselves, taking advantage of the "by prescription only" status of oral contraceptives. Physicians controlled not only pill prescriptions, but also knowledge about those pills, and many divulged only selected portions of that information "in the patient's best interest."

By 1970, fewer patients were willing to accept that kind of paternalistic treatment. As consumer advocates turned their attention to professional services, such as medicine, they criticized the unequal balance of power in the doctor-patient relationship.[56] Friction between doctors and female patients was further compounded by emerging feminist critiques of the male-dominated medical profession and of male-dominated society in general. The controversy over the safety of the pill offers a lens through which to view the increasing distance between patients, who demanded greater involvement in their treatment, and physicians, who endeavored to maintain their authority and control.

The Pill Hearings, Part II

The second round of hearings presented testimony from witnesses both in favor of and opposed to the birth control pill. Many witnesses devoted their testimony to bemoaning the results of the first round of hearings held a month earlier. These experts generally approved of the pill as a valuable means of fertility control, and they lamented the anxiety and alarm produced by the hearings and concomitant media coverage. Not coincidentally, several of the pill advocates were affiliated with large-scale family planning agencies, such as the Margaret Sanger Research Bureau, or population control organizations, such as the Population Crisis Committee and Planned Parenthood/World Population. One witness, Phyllis Piotrow, formerly of the Population Crisis Committee, anticipated that one hundred thousand unwanted "Nelson babies" would be born in the coming year as a result of women discontinuing the pill because of fear generated by the hearings. In an article for the *Progressive,* journalist Morton Mintz ridiculed this eponymy and its supporters (notably Senator Dole) by suggesting pill-related diseases be called "Piotrow strokes" or "Dole thromboembolisms."[57] The media immediately reported the concern about more unplanned pregnancies. In one of its articles on the Nelson hearings, *Time* printed a cartoon that showed one pregnant woman greeting another in an obstetrician's waiting room with the caption, "Corinne . . . You didn't tell me you gave up the pill, too?"[58]

Once again, Senator Nelson found himself in unexpected opposition to his usual allies. Nelson prided himself on being an environmentalist; he helped to originate the idea of Earth Day (which took place just two months after the pill hearings). As an environmentalist, he worried about the effects of overpopulation and supported most fertility control programs. However, his doubts about the safety of oral contraceptives and his leadership of the public hearings on the pill alienated him from the population control community. They felt that he had unnecessarily scared many women away from a very successful method of birth control. Nelson resented those accusations and defended his hearings in the interest of the public's right to know.

By the last days of the hearings, the central issue had boiled down to informed consent. Most physicians and scientists agreed that no new biomedical evidence had been presented to the Senate subcommittee; the debate over whether the pill caused cancer, for example, would have to wait for more data in order to be resolved. They disagreed, however, on how much of this information should be presented to patients. Some concurred with Nelson, who insisted that women should be given all available information about the pill so that they could make up their own minds. Others sided with the witness who testified: "To present the list of possible side effects as outlined in the present package insert to the average patient would serve no useful purpose, and would have many foreseeable and disastrous effects . . . A patient cannot reasonably be expected to make a profound professional judgment—she is not a doctor."[59]

The issue of informed consent in the use of oral contraceptives crystallized on the final day of the hearings, when FDA Commissioner Edwards announced that his agency planned to require pill manufacturers to include information for patients in every package of birth control pills. This pamphlet, written in lay language and directed to the patient, would outline the health risks associated with taking the medication. In his testimony, Dr. Edwards explained that the insert was "designed to reinforce the information provided the patient by her physician."[60] In the absence of good doctor–patient communication (which, according to the Newsweek-Gallup poll, characterized two out of every three women's experiences), the leaflet would supply the patient with the facts necessary to make an informed choice.

The hearings in February brought the first female witnesses to appear before the subcommittee. In total, four women testified: three physicians and the immediate past executive director of the Population Crisis Committee (Ms. Piotrow).[61] All of the women strongly advocated the use of the pill under a physician's supervision; none of them acknowledged the concerns expressed by the D.C. Women's Liberation protesters. However, it would be too simple to categorize these women as "anti-feminists"; in their minds, they were acting in the best interests of women in general. Of course, their conception of "women's best interest" was clearly colored by their socioeconomic position as upper middle-class professional women and varied considerably from the ideas of others. Dr. Elizabeth Connell, unlike many others, could afford the simultaneous luxury of a large family and a professional career. Near the end of her statement, she added, "I would like to make just a few comments on the part of the heretofore silent majority—women. The fact that I have chosen to pursue an active and exacting career while at the same time electing to produce six children should be my most potent evidence for my overwhelming belief in the larger role to which women should aspire."[62] Given Connell's generously sized family, her interest in overpopulation and pollution could be interpreted as classist. However, she also felt concern and sympathy for women with unwanted pregnancies, and annoyance with a system that did not allow women to make their own reproductive choices. Connell argued for legalized abortion and relaxed sterilization laws, as well as for the availability of a wide variety of contraceptive options, so that every individual could be free to decide how many babies to have and when to have them. Connell offered powerful rhetoric, in a style reminiscent of that of Margaret Sanger, to make her case for oral contraceptives:

> As a physician who began practice before the advent of the pill, I am constantly aware of the immense difference it has made to the lives of women, their families, and to society as a whole. The look of horror on the face of a 12-year-old girl when you confirm her fears of pregnancy, the sound of a woman's voice cursing her newborn and unwanted child as she lies on the delivery table, the helpless feeling that comes over you as you watch women die following criminal abortion, the hideous responsibility of informing a husband and children that their wife and mother has just died in childbirth—all these situations are deeply engraved in our memories, never to be forgotten. With the advent of more effective means of contraception, the recurrence of these nightmares was becoming blessedly less frequent.[63]

It would be inaccurate to dismiss Connell as insensitive to women. Her distance from the younger generation of activists pointed instead to their differing agendas, which precluded consensus on the best way to fight for women's reproductive rights.

In February 1970, D.C. Women's Liberation held a press conference to announce their intention to hold their own hearings on the pill. They specifically invited women who had suffered ill effects on the pill to testify. The women's hearings took place in March, three days after the Senate hearings ended, in a church in Washington, with child care provided. Flyers advertising the hearings read, "The Pill: Is It a Menace, A No-No, or a Girl's Best Friend?" About a hundred people turned out to hear testimony from Barbara Seaman; Sarah Lewit Tietze, a research associate at the Population Council (and wife of Dr. Christopher Tietze, also affiliated with the Population Council); Elaine Archer, a women's health care advocate; Etta Horn, a welfare rights leader, and others.

In their opening statement, the women of D.C. Liberation made their position clear: "We are not opposed to oral contraceptives for men or for women. We are opposed to unsafe contraceptives foisted on uninformed women for the profit of the drug and medical industries and for the convenience of men."[64] Their quarrel extended beyond the immediate problem of the safety of the birth control pill to encompass the entire existing system of traditional medicine: "It is not our mission to have all women on the pill discard it and change to another form of contraception. Our mission, if such it can be called, is to rise up, as women, and demand our human rights. We will no longer let doctors treat us as objects to be manipulated at will. Together we will ask for and demand explanations and humane treatment by our doctors and if they are too busy to give this to us we will insist that the medical profession must meet our needs. We will no longer tolerate intimidation by white coated gods, antiseptically directing our lives."[65]

Doctors did not have to be male to fit the model of "white coated gods." Women doctors, such as Elizabeth Connell, who did not disclose all information to patients about the side effects of the pill, were subjected to the same sort of criticism from the new health feminists. Connell set her sights on the more reformist goal of helping women to achieve reproductive control as a means to greater economic stability and personal fulfillment; she viewed the pill as a beneficial—but not perfect—technological solution to family planning. The members of D.C. Women's Liberation, by contrast, rejected reformist feminism because it did not address what they perceived as the underlying ills of a patriarchal system. Their objective was nothing less than the redistribution of power in society. In the realm of medicine, these feminists wanted women to participate as equal partners in their health care, which included the necessary prerogative of informed consent.

Thus, when FDA Commissioner Edwards announced the requirement for package inserts, many feminists regarded this as an important victory. After all, such an insert was a novel idea; oral contraceptives would be the first orally administered drug to carry a detailed warning directed at patients.[66] However, as feminists and other advocates of the package insert soon found out, the road from promise to reality was not easily navigated.

FDA and the Patient Package Insert

The 1938 federal Food, Drug, and Cosmetic Act obliged pharmaceutical manufacturers to make information about the safety of drugs available to physicians. In 1961, an amended version of the law required this information to be listed on the label of the package in the interest of "full disclosure," and within a few years most prescription drugs included a detailed package insert directed to physicians.[67] These pamphlets contained instructions for using the medications, as well as information on indications, contraindications, efficacy, and side effects. An editorial in the *Journal of the American Medical Association* pointed out that the drug companies composed and paid for the package inserts, thus making them little more than "promotional items."[68] Indeed, the FDA did not participate in the preparation of these pamphlets; the agency left decisions about their specific content to the manufacturers, so long as they met the basic requirements for drug labeling as outlined by the legislation.

The FDA regulations enacted to carry out the 1938 legislation created a new category of drugs available by prescription only.[69] While government regulation was designed to protect consumers from unscrupulous drug manufacturers, it also removed a significant amount of decision-making about medical treatment from the patient's, or consumer's, domain. After 1938, patients relied on physicians to instruct them on which drugs to purchase and to use. The doctor controlled not only the patient's treatment, but also the degree to which the patient understood the complexities of that treatment. Package inserts reinforced the physician's authority in medical matters. These pamphlets were intended for the eyes of physicians and pharmacists only; the prescribing physician transmitted as much or as little of the information to his patient as he thought necessary.

For more than thirty years, physicians had controlled the largely one-way flow of information about prescription drugs. Package inserts for patients would make them privy to at least some of these heretofore unavailable data. In the case of the oral contraceptives, women would be forewarned of possible side effects and health risks.

The day after the FDA Commissioner announced at the Nelson hearings his intention to require the insert, the *New York Times* published the proposed text of the pamphlet. Entitled "What You Should Know About Birth Control Pills," the 600-word document described in lay language the health risks, side effects, and contraindications of oral contraceptives. Although Edwards indicated that the insert was necessary because doctors did not adequately inform patients, the leaflet assured women of the competence of their doctors: "Your doctor has taken your medical history and has given you a careful physical examination. He has discussed with you the risks of oral contraceptives and has decided that you can take this drug safely."[70] Ten of the fifteen paragraphs in the proposed text referred to the doctor as the proper authority on oral contraceptives; the leaflet encouraged the woman to consult her physician in no fewer than six different situations.

In spite of this deference to the doctor, the medical profession strongly opposed the insert, claiming that the pamphlet would intrude upon the doctor-patient relationship. The pharmaceutical industry contended that the proposed leaflet overstated the potential risks and overlooked the benefits of oral contraception. Even the Department of Health, Education, and Welfare (HEW), which housed the Food and Drug Administration, argued that the leaflet needed to be revised to satisfy somewhat murky legal issues. (The *New York Times* reported that HEW was irked at having been left out of the loop on the development of the insert.[71])

In response to pressure from professional, industrial, and government interests, the FDA backed away from its initial proposal and substituted a much shorter, less detailed version. The revised text, 100 words in length, mentioned only one complication of oral contraception, blood-clotting disorders. Whereas the first draft had included statistics on increased risk and mortality rates from thromboembolism, the edited version omitted this information. It encouraged women to see their doctors if they experienced side effects, listing just five symptoms and conditions whereas the earlier draft had listed more than twenty-five.[72]

Outraged by this turn of events, women from D.C. Women's Liberation staged a sit-in at the office of HEW Secretary Robert Finch to protest the watered-down version.[73] Although she was not present at the sit-in, Barbara Seaman recalled what had happened: "It was a real 1960s sit-in. These women came with their little children, and they sang. They made up pill songs to the tunes of nursery rhymes, like 'The doctors give the pill, the doctors give the pill, heigh ho the derry-o, the women take the pill, the women they get ill, the doctors send their bill.'"[74] Alice Wolfson, organizer of the demonstration, also remembered the sit-in: "You've got to remember the times. They were so different. We were really into pushing everything, so we brought babies, those of us who had babies . . . we brought them everywhere, and when anyone would say, 'take the kids out of the hearing room' or this or that, we would say, 'well, that's fine, where's the day care center?' All of these issues were always linked . . . we were really militant."[75]

Secretary Finch did not see the feminists that day, but agreed to meet with them a few days later. On March 30, 1970, Secretary Finch, FDA Commissioner Edwards, and Surgeon-General Jesse Steinfeld met with six representatives of the feminist group, including Barbara Seaman and Alice Wolfson. Wolfson reported in the feminist newspaper *off our backs* that after an hour of discussion, "Finch stormed out of the room, tailed by Edwards and Steinfeld, claiming we were just wasting his time."[76]

The feminist petition to reinstate the stronger version of the package insert did not sway the officials of HEW and FDA. On April 10, 1970, the FDA published the abridged draft of the oral contraceptive package insert in the *Federal Register* and invited all interested parties to respond with comments. During the next thirty days, letters from more than eight hundred individuals and groups flooded the offices of Secretary Finch, Commissioner Edwards, and the Hearing Clerk at the FDA in Rockville, Maryland.

Of these eight hundred responses, about a third requested more information about the insert and about the oral contraceptives in general. Many had not read the actual notice in the *Federal Register*, but had heard or read about it in the news. More than half wrote to object to the abridgement of the proposed text. Most of these were copies of form letters distributed by women's and consumers' groups which complained that the insert did not provide full disclosure on the adverse effects of the pill and called for public hearings on the matter. Over a hundred women and men wrote their own letters to protest against the reduction in the length and scope of the insert. Some of them objected to the unequal distribution of power in the doctor-patient relationship:

> I have inadvertently received the physician's copy of facts and cautions now included in each 3-pack of Ortho-Novum pills; the first time I told the people who had given it to me and they said "You're not supposed to see that. That's only for doctors." I was outraged and insulted at this; the only reason I can see for doctors or other parties to withhold medical information from patients is the desire to maintain their psychological and monetary power over us.[77]

Others added their concern about the integrity of the pharmaceutical industry and its control over government agencies:

> In the *Chicago Daily News*, March 24, 1970, I read that the FDA has called for toning-down the wording of the precautionary literature on oral contraceptives which HEW has in the planning stage. It was clear from the article the HEW is bowing to pressure from the drug industry, the AMA, and many private physicians, all of whom feel that a precise report will harm the Pill's market and their own pocketbooks.[78]

> I request a public hearing on this urgent matter. A matter which has caused the consumer again to believe that government and Industry will gladly compromise our health in the name of advancement.[79]

Still others expressed the opinion that women had the right to full disclosure on medical matters: "I DEMAND,—that as a woman, having the option to take the pill or not, I have *all* facts in front of me!"[80]

Much of the public interest in the oral contraceptives can be attributed to the publicity generated by the Nelson hearings and the consistent coverage by the news media of the controversy over the safety of the pill. It is less easy to explain why people moved beyond mere interest to direct action—in this case, writing letters of protest. Most likely the climate of the times spurred many individuals to action. Within a society attuned to the issue of rights and conducive to political activism, people felt empowered to speak out against what they perceived as a denial of the right to informed consent. During this time, many Americans felt angered by the secret policies of the government in the Vietnam War; by 1970, the demand for public information extended to a broad range of government activities. In addition, Congress had recently passed the Freedom of Information Act (in 1967), which both entitled and emboldened citizens to seek information previously withheld from them. The demand from hundreds of people for a public hearing on the content of the patient information insert in 1970 fit into this larger social context.

It is interesting that the Planned Parenthood Federation did not take a position on the proposed patient package insert. In early March, the medical director wrote a polite letter to Commissioner Edwards on behalf of the organization to request more information about the exact nature and implications of a patient package insert. However, Planned Parenthood did not send a statement in response to the notice in the *Federal Register*.[81] A few local family planning organizations did send letters to the FDA. In general, these groups argued that the insert be reworded so as not to scare women away from the pill. They did not want the package insert to interfere with the high level of patient acceptance of oral contraceptives at birth control clinics.[82] Later, after the FDA commissioner had made his final decision on the insert, *Newsweek* counted Planned Parenthood among the opponents of a detailed warning, citing concerns that such a warning would drive women away from birth control clinics and would interfere with the doctor–patient relationship.[83]

Doctors also wrote to the FDA. The majority opinion of the medical profession was represented in letters from the American Medical Association, the American College of Obstetricians and Gynecologists, the Association of American Physicians and Surgeons, the American Society of Internal Medicine, the AMA Interspecialty Committee (representing nineteen medical specialties), the California Medical Association, the Rhode Island Medical Society, the Texas Medical Association, the South Georgia Medical Society, and the Alameda–Contra Costa Medical Association of California.

With very few exceptions, doctors strongly opposed the patient package insert. Their objections fell into two main categories: the insert would interfere with the doctor-patient relationship, and the government should not regulate what information the doctor must give to each patient. Excerpts from physicians' letters reveal their indignation at regulation from outside the medical profession:

> I deeply resent the Government of the United States coming between me and my patients in the matter of a single class of prescription items . . .[84]

> This practice can serve no useful purpose accept [sic] to frighten patients. Furthermore, it represents an undesirable entry of government into the practice of medicine.[85]

> Although I believe firmly in the principle of informed consent with regard to doctor–patient relationships, I must point out that the FDA by this blandly, intrusive approach is apparently bypassing the individual physician completely and thus suggests its lack of confidence in the entire medical profession[']s competence and ethics.[86]

> I would sincerely hope that before such a proposal becomes an accomplished fact, strong consideration be given to the far-reaching [e]ffects such legislation [regula-

tion] will have on the medical community of this country, which has been so frequently maligned . . . The determination of appropriate use of medications must continue to rest in the hands of the physician . . . To remove this clinical relationship would be just another method of eroding the foundation of American medicine.[87]

A few physicians expressed their approval of the idea of a patient package insert for oral contraceptives. These doctors agreed with consumer advocates that the patient should be fully informed before making the decision to use birth control pills. However, their position contrasted sharply with the large majority of physicians, who preferred to retain the prerogative of how much and what kind of information to give to each patient.

One doctor who wrote to the FDA in support of government involvement in disseminating information about prescription drugs argued that the patient package insert would "serve as a protection for the doctor rather than as a cause for initiating lawsuits."[88] One would think that this reasoning would appeal to the pharmaceutical industry as well; by 1970 more than a hundred lawsuits had been filed against birth control pill manufacturers.[89] However, drug companies vehemently opposed the inclusion of an FDA-mandated warning in packages of oral contraceptives. The president of the Pharmaceutical Manufacturers Association (PMA), which represented 125 drug companies, articulated the industry's objections, both general and specific, to the proposed insert; four companies also wrote letters supporting the PMA position and offering additional comments on the proposed wording.

The manufacturers pointed out that the insert contradicted the intent of the federal Food, Drug, and Cosmetic Act, which clearly distinguished between prescription and over-the-counter drugs. They argued that the act designated the class of prescription drugs to be issued on a physician's prescription only and thus precluded the necessity of directing detailed information to the patient. Furthermore, "the complexity of prescription drugs as well as the delicate nature of the physician-patient relationship requires individual decisions in each case as to what information should be imparted to a patient and as to how it should be conveyed."[90] The drug industry sided with the medical profession in preserving the sanctity of the doctor-patient relationship.

The PMA also objected to several specific points in the proposed insert. If in fact the FDA did decide to require such an insert, then the manufacturers wanted to address several errors of omission and commission. First, the organization contended that the warning about oral contraceptives ought to be balanced by a statement of the health risks associated with pregnancy. Second, they rejected the statement that read, "these points were discussed with you when you chose this method of contraception," because the manufacturer could not presume to know what every physician told every patient when prescribing birth control pills. Third, they disputed the declaration that blood clotting disorders were a "known complication" of oral contraception, preferring instead to describe the relationship between the pill and thromboembolism as merely "associated." Fourth, they objected to the list of five symptoms because (a) the power of suggestion might cause patients to imagine side effects, and (b) the list omitted other side effects associated with oral contraceptives.[91] The PMA stated the concern that patients might not report other symptoms; an unstated concern may have been the fear of litigation if a woman experienced conditions not specified in the insert.

The content of the insert that became a requirement in June 1970 was significantly modified from the version proposed in April. The *New York Times* reported that the change resulted from pressure from physicians; clearly, the FDA bowed also to the interests of the powerful pharmaceutical industry.[92] Although it did describe abnormal blood clotting as "the most serious known side effect," the insert did not list any symptoms; instead, it told the reader to "notify your doctor if you notice any physical discomfort."[93] Four of the seven sentences in the insert described the availability of an information booklet, which the patient could request from the physician.

This 800-word booklet was written by the American Medical Association in conjunction with the FDA and the American College of Obstetricians and Gynecologists; it resembled in scope and content the original insert proposed by the FDA back in January. The package insert merely informed the patient that further information was available; the onus fell on the patient to ask her doctor to give her the booklet. Commissioner Edwards felt that "the prescribing physician should be the person to provide his patient with the necessary information to assure her safe use of the prescribed medication"; the insert served to remind the patient (and, it was hoped, the physician) that "careful doctor–patient discussion about the use of the drugs" should take place at regular intervals.[94] In this way, Edwards aimed to appease the consumers, feminists, and patients who demanded informed consent, and the physicians and manufacturers who wanted control to remain in the hands of the medical profession.

Of course, neither group was wholly satisfied. Consumers and feminists objected to the conscious withholding of information from patients; the FDA insert-booklet compromise allowed physicians to choose not to tell some patients the whole story. In late June, a woman announced that she planned to sue the FDA and HEW under the 1967 Freedom of Information Act for disclosure of all clinical and toxicological records on eight different brands of oral contraceptives.[95] According to HEW, release of the data from laboratory and clinical tests "would only confuse the average citizen."[96] The plaintiff sought to challenge that assumption.

In another lawsuit, an associate of the consumer advocate Ralph Nader lost a battle to have the 800-word booklet included directly in packages of birth control pills in lieu of the shorter insert.[97] Barbara Seaman followed the development of the patient package insert closely. She recalled: "We were very disappointed to learn that a compromise had been reached. We were very shocked when we learned that the compromise was that it was going to be distributed to doctors' offices, not by pharmacists, and that the prescribing physician would give out the booklet . . . I spent a lot of time buttonholing women and asking them if they had received this, and I invariably found that if they'd been to a clinic, the chances were very good they had gotten it. Private physicians, forget it! I don't know if I ever found one person whose private physician gave it to her."[98]

The battle over the patient package insert demonstrated the influence and power of the medical profession and the pharmaceutical industry over the FDA. Commissioner Edwards yielded twice to the demands of these two powerful institutions: first, in cutting the insert text to one-sixth of the original proposed before publication in the *Federal Register,* and second, in further reducing the strength of the final warning. After more than two years of Senate inquiry into the drug industry, pharmaceutical manufacturers still held the upper hand in the uneasy relationship with their regulatory agency, the FDA.

Given its subordinate position to the industry it was supposed to regulate, why did the FDA bother to suggest the insert in the first place? Three factors contributed to this action. First, Charles Edwards, who had recently replaced the rather ineffectual Herbert Ley as commissioner of the FDA, wanted to improve the status of his agency. A patient package insert might help to rein in the powerful drug industry.[99] Second, the FDA felt pressure from legislators—Senator Nelson, in particular—to address the problem of the oral contraceptives. Since the manufacturers would not voluntarily provide information to consumers, the FDA had to respond to the demand of the public and its elected officials for action. Third, the climate of skepticism toward medicine and big business by 1970 seemed favorable to regulation of the profession and its commercial cousin, the drug industry.

In spite of its watered-down wording, the insert still represented an important turning point in the doctor-patient relationship. Patients had demanded the right to know about the medications prescribed for them, and the package insert legitimized this claim.

Eight years later the FDA ordered the minimal patient package insert that accompanied oral contraceptives to be replaced with a lengthy information leaflet.

In 1977, the FDA issued patient information requirements for estrogens used in hormone replacement therapy. In the wake of this mandate, the FDA decided to revise the oral contraceptive requirements to be consistent with those of other estrogen products.[100] This 1978 version of the patient package insert repeated the information contained in the physician package insert in lay language in simplified form, and represented what consumer and feminist groups had wanted all along.

As consumer advocates fought to extend inserts for patients to other prescription drugs, many physicians viewed the new developments with dismay. An editorial in the *Journal of the American Medical Association* complained: "The FDA is responding to pressure from legislative and consumer groups who are demanding that patients have 'a right to know' more about medication doctors prescribe . . . The final decision to give a PPI to a particular patient should be left to the physician and not legislated bureaucratically as a response to pressure groups."[101] Physicians regarded the patient package inserts as serious encroachments onto their professional territory. The FDA-mandated insert represented a direct affront to the autonomy that the profession had worked so hard to maintain throughout the twentieth century.[102]

Nevertheless, in the early 1970s the patient package insert for oral contraceptives had minimal impact on the interactions between doctors and patients. Doctors continued to assert their authority and control over patients. For example, in an article written in 1970 on the use of oral contraceptives to manage postpartum breast engorgement and discomfort, the authors extolled the benefits of oral contraception for women of low socioeconomic status and commended a dominant role for the physician in contraceptive decision-making: "We submit that if we seriously hope to meet the plight of the already overburdened, underprivileged, lower socioeconomic group, we can do no better than extend this technique [immediate postpartum oral contraception] to all clinic services on a nationwide basis. Only here will be found such a selective, receptive, and at least semi-captive group. Lack of incentive, a previous enigma, is now replaced by physician incentive and control."[103] Despite the pretense of objectivity and impartiality associated with a scientific publication, the authors felt that such editorial comment was legitimate. It was precisely this kind of paternalistic attitude that so infuriated the new health feminists and that served as a focal point for their movement.

The Women's Health Movement

The controversy over the safety of the pill and the subsequent debate about package inserts for patients spurred feminists to action.[104] In her study of the women's health movement, Sheryl Burt Ruzek noted a transformation in the self-images of women who took part in the general feminist movement of the late 1960s. These women used their newfound self-confidence and assertiveness to challenge the practices and assumptions of the traditional, male-dominated medical system that had been in place for decades.[105] They did not recommend that the pill be banned; as Barbara Seaman said, "you can't put the genie back in the bottle."[106] Instead, they championed women's right to full disclosure and informed consent, not only in birth control but in all drug therapies and medical treatments.

Oral contraception was not the first reproductive health issue to engage women activists in the decades after World War II. The La Leche League began to promote breastfeeding as a better alternative to infant formula in 1956. This group challenged the authority of "scientific motherhood" and advocated the return of mothering to the mother.[107] At the same time, women dissatisfied with the anesthetized experience of hospital childbirth turned to the "natural childbirth" techniques of Grantly Dick-Read.[108] A third early feminist health issue centered on the promotion of physical therapy for radical mastectomy patients to prevent the loss of mobility in their arms.[109] All of these initiatives sought to redress problems created by the medicalization of women's health in the twentieth century.

By the late 1960s, women around the country gathered in informal "consciousness raising" groups to share their experiences as women in a sexist society. Many of these groups focused their discussions on medicine, health, and the body. In Boston, a group of women frustrated with the existing system of medical care decided to do research on health-related topics; the resulting papers written by members of this Boston Women's Health Book Collective formed the basis for the bestselling *Our Bodies, Ourselves*.

A rallying point for the early health feminists was the legalization of abortion. Members of D.C. Women's Liberation became more interested in women's health issues as they fought for legalized abortion. Alice Wolfson recalled: "We found that one of the leading causes of maternal death in D.C. General Hospital, which is the only public hospital in D.C., was botched abortions. So when we began to scratch that surface, we began to find out lots of other things about health care for women, and also began to develop a kind of analysis of the health care system and the relationship of women to power and the health care system as being kind of a microcosm of women's place in society in general at the time."[110] It was only a short step from activism on abortion and sterilization abuse to activism on a broader range of health-related issues. In 1970, many abortion rights advocates were still working at the state level. The pill hearings, however, took place on Capitol Hill and attracted national attention.

At the Nelson hearings, Barbara Seaman met Alice Wolfson, thus bringing together the "uptown" and the "downtown" feminists on the issue of the safety of the pill. Seaman used the term "downtown" to describe the more radical feminists who belonged to militant groups such as the Redstockings, Bread and Roses, and D.C. Women's Liberation. "Uptown" referred to those feminists who belonged to more mainstream organizations such as Betty Friedan's National Organization for Women. Seaman described herself as an uptown feminist . . . until she met Wolfson: "I instantly struck up a friendship with Alice. I was so impressed with her. I just thought everything she said was right, putting her political analysis on all of this. I thought, yes, yes, why didn't I see it this way, why didn't I understand it this way all along? But of course, if I had seen it that way, I wouldn't have been suitable to write for the *Ladies' Home Journal* or to write the kind of book I wrote. So in the scheme of things, it's just as well that I saw it from my particular lens up until that moment in January, 1970, when I met Alice."[111]

The two women remained in close contact after the hearings; Wolfson lived in Washington and could keep track of the developments at the FDA. Soon thereafter, she invited Seaman to testify at the women's hearings on the pill, to meet with HEW Secretary Finch, and to "sit in" at a closed meeting of the FDA's Advisory Committee on Obstetrics and Gynecology: "Alice and I went down there [to the FDA] and we walked in [to the room where the Advisory Committee met] and we sat down and everybody said to us, "What are you doing here?" And Alice said, "Well, why shouldn't we stay? It's our bodies you're discussing." They told us we had to go out, and they would discuss whether we would be allowed to stay. I remember we sat on the floor in the hallway right outside, and I think we were there about a half an hour, and then somebody came out and said okay, we could stay . . . I think that was the first time that the advisory committees were made open to the public."[112] For both women, full disclosure and informed consent remained at the heart of all of their battles.

Wolfson and Seaman were in the vanguard of a branch of the more general women's movement that was to become the women's health movement. In 1971, D.C. Women's Liberation broke up when lesbian and straight members could not agree on the mission of the movement. Wolfson, who was pregnant at the time, could not abide the radical lesbian position against all men, and turned instead to health feminism because of its focus on a single group of issues. In 1975, she and Seaman joined with other health activists to form the National Women's Health Network. The network acted as an information clearinghouse for women's health collectives around the country. It represented women's health interests in hearings held by Congress and the FDA and published a newsletter on health policy and legislation.[113] Twenty years later, the organization remains actively engaged in women's health issues.

The goals of health feminists in the 1970s differed dramatically from those of their predecessors. A generation earlier, Margaret Sanger and Katherine McCormick lobbied for women's right to reproductive control. To achieve their goal, they enlisted the help of scientists and physicians and encouraged these experts to develop a solution to the problem of birth control. Sanger and Mc-Cormick hailed the oral contraceptive as a scientific triumph for women and gladly entrusted contraception to the hands of physicians. In contrast, the women's health movement rejected the hegemony of the medical-pharmaceutical complex and instead advocated lay control over the delivery of health services.[114] Health feminists objected to the birth control pill on several grounds: insufficient clinical trials, potentially fatal side effects, and a lack of informed consent among its millions of users worldwide. In 1970, feminists interpreted the pill as representative of patriarchal control over women's lives; it was this issue that catalyzed the rise of the women's health movement.

Although not as well organized nor as powerful as the established medical profession and the pharmaceutical industry, health feminists were determined to take on these male-dominated institutions and their traditional assumptions and practices. In the decades to follow, the interests of feminists, female patients, physicians, drug manufacturers, and government officials would clash many times over issues such as diethylstilbestrol (DES), intrauterine devices, Depo-Provera, Norplant, and abortion. All of these debates had their own unique set of concerns; however, in each one the matter of informed consent, as articulated in the controversy over the safety of oral contraceptives, remained central.

Chapter Six

Conclusion

For historians of the 1950s and 1960s, the pill serves as a barometer of changes in attitudes toward science, technology, and medicine. At the same time that eager acceptance of the pill gave way to caution and concern, trust and confidence in medical research and its products also yielded to questioning and uncertainty. However, concern about the safety of the pill did not lead to its wholesale abandonment. In a similar fashion, broader questions about the practice of medicine and medical research did not result in an overhaul of the existing system or in rejection of the applications of medical science. As a specimen of medical technology, the pill offered powerful benefits to outweigh its risks. In the eyes of many contemporaries, science and medicine made significant contributions to the quality of life, which overshadowed their negative aspects. Although Americans might have expressed skepticism toward medical science and its products, for example, the pill, they continued to embrace the culture of "modern" medicine and technology after the 1960s.

By 1990, 80 percent of all American women born since 1945 had used the pill at some time in their lives.[1] A series of national fertility studies confirmed that more women used the pill than any other reversible method of birth control through the 1970s and 1980s. However, the number of women on the pill fluctuated greatly during this period. The percentage of married women taking the pill dropped from 36 percent in 1973 to 20 percent in 1982. Many of these women opt-ed for sterilization; between 1973 and 1982, the rate of female sterilization among married women aged fifteen to forty-four more than doubled (from 12% to 26%). More married men also underwent vasectomies, so that by 1982 15 percent of those aged fifteen to forty-four had been sterilized.[2] By 1988, almost half of married couples relied on either male or female sterilization to avoid pregnancy.[3]

The decline in the percentage of pill users can be attributed in large part to repeated "health scares" about oral contraceptives. In 1975, for example, the media reported a medical study linking the pill with an increased risk of stroke, prompting many women to discontinue use of the pill.[4] In spite of medical reports that oral contraceptives also protected against uterine and ovarian cancers, consistent coverage of the problematic association between oral contraceptives and breast cancer cast a pall over the pill. Although a 1996 analysis of 54 epidemiological studies involving more than 150,000 women concluded that birth control pills did not increase the risk of breast cancer, it is unlikely that this report represents the final word on the subject.[5]

The medical controversy over the safety of the pill that began in the 1960s was never resolved. A scientific dispute resists resolution when opposing experts belong to different scientific communities with different perceptions of the evidence at hand.[6] In the case of the pill, individual physicians and medical researchers assessed the available information on adverse health effects in different ways, depending on their own experience with patients and the results of their own research. In the early 1970s, the pill controversy died out instead of achieving closure through consensus, sound argument, or negotiation. In spite of all the negative publicity, women continued to use the pill, and doctors continued to prescribe it for their patients. Medical researchers and epidemiologists carried out studies of the long-term health effects of oral contraception; although news of their work caused the debate to flare up on a number of occasions over the next two decades, none of it seriously detracted from the pill's enduring popularity.[7] Health feminists shifted their attention to the more pressing matters of abortion, DES, and the IUD, where activism could achieve results. Beyond the patient package insert, there was little else to do about the pill.

In parallel with the efforts of the patients' rights movement (an offshoot of the consumer movement), the women's health movement impelled the medical profession to seriously consider the issue of informed consent. As early as 1969, the California Medical Association encouraged its members to distribute booklets entitled "What You Should Know about the 'Pill'" and "How Does Your Doctor Know When a New Drug Is Safe?" to all women who requested pill prescriptions. An article in *California Medicine* advised physicians to "make an entry in the patient's record of the delivery of the articles. In this fashion, 'informed consent' could occur without the physician being required to be a lecturer but with the explanation to the patients being uniform and being done in a friendly, understandable manner . . . The importance of an entry in the patient's record of whatever information you give cannot be overemphasized."[8]

In 1970, in the wake of the Nelson hearings, the American Medical Association counseled physicians to give their patients information about the advantages and disadvantages of the pill and other methods of birth control "to protect themselves against possible malpractice suits." The AMA echoed the *California Medicine* article's suggestion to note in the patient's medical history that the side effects of different contraceptives had been discussed.[9] From this perspective, informed consent served not to help the patient make an educated decision but rather to cover the prescribing physician in case the patient sued.

The advent of the patient package insert released physicians from liability to a certain extent. In theory, if not in actual practice, the patient package insert provided women with information on which to base their contraceptive decisions. Nonetheless, physicians resented the insert as an intrusion of the government into the practice of medicine. Patients, by contrast, welcomed the package insert, which promised to offer them direct access to medical information. Through the 1970s, the insert did not in fact provide much useful information but rather served as a symbolic reminder of the patient's right to informed consent.

In the 1970s, health feminists became even more disillusioned with the existing system of women's health care. Barbara Seaman wrote two more books, *Free and Female* and *Women and the Crisis in Sex Hormones,* to complete her "estrogen trilogy," before moving on to a biography of the author Jacqueline Susann. Other feminist writers turned their attention to other issues of reproductive health, spawning a whole new field of literature that included not only critiques of patriarchal medicine but also self-help books on subjects such as premenstrual syndrome, menopause, pregnancy, and childbirth.[10] The women's health movement sought to increase women's options in medical care. While many women did take advantage of alternative services, the majority continued to go to traditional (usually male) physicians for medical services, including birth control.

At the same time, feminists also began to criticize the population control movement for its insensitivity to individual needs and its palliative approach to problems requiring more complex social and economic solutions. In her 1976 book *Woman's Body, Woman's Right,* Linda Gordon exposed the motives and methods of advocates of population control as counter to the needs and wishes of poor, nonwhite, and Third World women.[11] During the 1960s, these advocates first based their policies on the pill on cost. Population control agencies, such as the Population Council, sought an inexpensive contraceptive to export to rapidly growing Third World countries; when the pill first came out, they dismissed it as too expensive for anyone but the wealthiest to afford. Instead, they focused their energies on the development of the IUD, an inexpensive, low-maintenance method of birth control. Later in the decade, the population control position shifted in response to the emerging debate over the safety of the pill. As reports of the health risks of the pill began to surface, population control advocates rearticulated their recommendations: middle-class and better educated women should use the diaphragm and only those who "couldn't" or "wouldn't" use the safer barrier methods should continue to use the pill.[12] Health feminists objected to the classism in these recommendations, which implied that in spite of its adverse health effects, the pill was "good enough" for poor women who needed to control their fertility.

Part of the health feminists' concern focused on the pill as a technological "fix" to the complex social problems of overpopulation and poverty. Controlling the fertility of women, they argued, was more than a matter of providing access to contraception; it required that the social, cultural, educational, and economic situation of women be addressed as well.[13] This issue dovetailed with the reaction of the women's health movement against the excessive use of technology in medicine. Health feminists sought to provide more "natural," woman-centered practices into health care. They advocated the diaphragm and the cervical cap—both barrier contraceptive methods controlled by women—over the pill and IUD. How surprised Margaret Sanger (who died in 1966) and Katherine McCormick (who died in 1967) would have been at this turn of events!

In spite of the backlash against the pill at the end of the 1960s and continued doubts about its safety through the 1980s, the birth control pill has had an undeniably significant effect on certain segments of the American population. White, middle-class, married women, who could afford the pills and the private physicians who prescribed them, benefited most from oral contraception in the 1960s. The pill offered easy, reliable, effective protection against pregnancy, which empowered women to plan when to have children and how many to have. Earlier methods of birth control had also enabled women to plan their families, but with less certainty and less success than the pill. This freedom in turn contributed to women's increasing strides in developing careers outside the home. Of course, numerous other sociocultural barriers to women's employment had to fall before women could choose their careers freely, but the ability to delay and to space childbearing helped to further trends in women's work roles.

Unmarried women did not really begin to take advantage of the pill until the 1970s. The author of a recent study of the American family in the twentieth century identified the 1970s as the "high point of the sexual revolution."[14] She cited five social and demographic changes, in addition to the development of the pill, important to the expansion of sexual liberalization: the rising age of marriage, equality of education for men and women, increased autonomy for women, the increase in the number of single people as the baby boom generation came of age, and the reaction of the baby boom generation against the perceived "hypocrisy of their elders." Many of those trends had begun in the 1960s and continued to expand in the following decade. In the 1970s, fewer people felt obliged to wait until marriage. Sex became "democratized" as premarital sex spread to the mainstream population. More single women began to use contraception (mainly the pill) in the 1970s, especially after the Supreme Court's decision in *Eisenstadt v. Baird* in 1972 effectively extended the right to birth control to everyone, regardless of marital status. Between 1970 and 1975, the percentage of college women reporting that they had had sexual intercourse jumped twenty points, from 37 percent to 57 percent.[15]

In the 1980s, the advent of AIDS forced women and men to reconsider their sexual activities and their birth control choices. The pill offers no protection against the HIV virus or any other sexually transmitted disease; only the condom affords such protection. For many women, fear of acquiring the HIV virus replaced the fear of getting pregnant, and the confidence they placed in the pill no longer sufficed. Although nineteen million American women still rely on the pill for birth control, an unknown proportion of these women either double up with condoms or take their chances.

In the early 1990s, two new contraceptive methods became available in the United States, neither of which offered any protection against sexually transmitted disease.[16] Norplant, a subdermal implant that continuously releases a synthetic hormone into the blood over a five-year period, has been embroiled in social controversy because of legislative proposals offering incentives to welfare mothers who use it and judicial decisions mandating its use in the sentencing of abusive mothers.[17] Depo-Provera, a hormone injection whose contraceptive effect lasts for three months, was developed in the 1960s, but did not receive FDA approval until late 1992 because of concerns about its potential carcinogenicity.[18] Feminist groups object to both Norplant and Depo-Provera because these long-lasting contraceptives may facilitate introduction of voluntary or mandatory birth control programs and because reliance on these hormonal methods may curtail the use of condoms, thereby increasing the risk of exposure to sexually transmitted diseases. By the late 1980s, all but one of the major pharmaceutical companies in the United States had withdrawn from the field of contraceptive research and development, owing perhaps as much to the increasingly litigious nature of American society as to the extremely high cost of birth control development and testing.[19] As a result, further innovations in contraception are unlikely to be realized in the near future.

So the pill, nearing its fortieth birthday, is still going strong. Public concern about its adverse health effects has died down or is at least in a state of quiescence until the next medical report appears. In spite of its lack of protection against the HIV virus, the pill remains the most popular reversible method of birth control among American women; given the trend in contraceptive use, it seems unlikely to lose its first-place ranking to another method. Recently, the U.S. Food and Drug Administration has considered proposals to change the pill from its prescription-only status to an over-the-counter drug.[20] Those in favor of this change claim that neither safety nor efficacy issues warrant prescription status for the pill. An over-the-counter birth control pill might reach many more women than the current prescription pill; furthermore, it would return contraceptive choice to women, without the physician as intermediary. Opponents warn that if the pill lost its prescription status, Medicaid, Planned Parenthood, and other clinics might not be able to offer free or reduced-price oral contraceptives, thus denying access to low-income women. They also argue that women who bought their pills at the drugstore would have little incentive to go to doctors or clinics for preventive medical care (e.g., Pap smears and breast exams) and that without adequate medical supervision, women might not learn to use the pill safely or reliably.

Forty years after appropriating control over contraceptive services, physicians still maintain authority in the realm of birth control. For now, the pill remains firmly within the jurisdiction of the medical profession. Whether it becomes an over-the-counter item, sold alongside condoms and spermicidal inserts, or continues as a prescription drug, available only after a physician visit, women will continue to take the pill. In the final analysis, the pill's legacy to women is the belief in and the right to simple, safe, and reliable contraception.

Notes

Introduction

1. Ashley Montagu, "The Pill, the Sexual Revolution, and the Schools," *Phi Delta Kappan* 49 (May 1968): 480.

2. "The Age of the Thing," *Economist*, December 25, 1993, 48.

3. Carl Djerassi, "The Making of the Pill," *Science 84*, November 1984, 127.

4. Rosalind Pollack Petchesky, *Abortion and Woman's Choice: The State, Sexuality, and Reproductive Freedom* (New York: Longman, 1984), 169–76; Alexandra Dundas Todd, *Intimate Adversaries: Cultural Conflict between Doctors and Women Patients* (Philadelphia: University of Pennsylvania Press, 1989), 30–31, 77–98; David J. Rothman, *Strangers at the Bedside: A History of How Law and Bioethics Transformed Medical Decision Making* (New York: Basic Books, 1991), 142; Paul Starr, *The Social Transformation of American Medicine* (New York: Basic Books, 1982), 380.

5. For an introduction to population control, see Linda Gordon, *Woman's Body, Woman's Right: Birth Control in America* (New York: Penguin, 1990), 386–96. For more detailed investigations of specific population control programs, see Annette B. Ramirez and Conrad Seipp, *Colonialism, Catholicism, and Contraception: A History of Birth Control in Puerto Rico* (Chapel Hill: University of North Carolina Press, 1983), and Martha C. Ward, *Poor Women, Powerful Men: America's Great Experiment on Family Planning* (Boulder, Colo.: Westview Press, 1986).

6. Gordon, *Woman's Body, Woman's Right*, xviii–xix.

7. For participants' accounts, see Egon Diczfalusy, "Gregory Pincus and Steroidal Contraception: A New Departure in the History of Mankind," *Journal of Steroid Biochemistry* 11 (1979): 3–11; Victor A. Drill, "History of the First Oral Contraceptive," *Journal of Toxicology and Environmental Health* 3 (1977): 133–38; Joseph W. Goldzieher and Harry W. Rudel, "How the Oral Contraceptives Came to be Developed," *Journal of the American Medical Association* 230 (October 1974): 421–425. For journalists' accounts, see Paul Vaughan, *The Pill on Trial* (New York: Coward-McCann, 1970); Loretta McLaughlin, *The Pill, John Rock, and the Church* (Boston: Little, Brown, 1982).

See also IIT Research Institute, *Technology in Retrospect and Critical Events in Science* (Chicago: Illinois Institute of Technology, 1968), a report commissioned by the National Science Foundation in 1967 and published in 1968. This project consisted of the "retrospective tracing of key events" leading to five important technological innovations—magnetic ferrites, video tape recorder, electron microscope, matrix isolation, and the oral contraceptive pill—in order to elucidate the relationships among non-mission research, mission-oriented research, and development in the production of new technology. The report concluded that all of the technological innovations resulted from nonmission-oriented research (not a shocking finding, especially given the report's sponsor). In the case of the oral contraceptive, almost all research prior to the 1950s was categorized as non-mission research, while the work done in the ensuing years was labeled mission-oriented research. Once the "mission" had been articulated by its participants, then obviously their research was directed toward the achievement of that goal. The final report was published in 1973 and included five new case studies in addition to further research on three of the original innovations, one of which was the oral contraceptive pill. See Battelle, *Interactions of Science and Technology in the Innovative Process: Some Case Studies* (Columbus, Ohio, 1973).

8. Gena Corea, *The Hidden Malpractice: How American Medicine Mistreats Women* (New York: Harper Colophon, 1977, 1985), 151. For another critique of the development of the pill as inimical to the needs of women, see Kristin Luker, *Taking Chances: Abortion and the Decision Not to Contracept* (Berkeley: University of California Press, 1975), 124–26.

9. Rosalind Pollack Petchesky, *Abortion and Woman's Choice: The State, Sexuality, and Reproductive Freedom* (New York: Longman, 1984), 171.

10. Catherine Kohler Riessman, "Women and Medicalization: A New Perspective," *Social Policy* 14 (1983): 3. Adele Clarke discusses Riessman's work as a significant step forward in feminist analysis in "Women's Health: Life-Cycle Issues," in *Women, Health, and Medicine in America: A Historical Handbook*, ed. Rima D. Apple (New Brunswick, N.J.: Rutgers University Press, 1992), 8–10.

11. Judith A. McGaw, "Review: Women and the History of American Technology," *Signs* 7 (Summer 1982): 803. For another view of the forces shaping the development of contraceptive technologies and the consequences of using those technologies, see Judy Wajcman, *Feminism Confronts Technology* (University Park: Pennsylvania State University Press, 1991), 74–78.

12. For an interesting study of the popularization of science in mass-circulation magazines, see Marcel C. LaFollette, *Making Science Our Own: Public Images of Science, 1910–1955* (Chicago: University of Chicago Press, 1990). See also Dorothy Nelkin, *Selling Science: How the Press Covers Science and Technology* (New York: W. H. Freeman, 1987) and Hillier Krieghbaum, *Science and the Mass Media* (New York: New York University Press, 1967).

13. More than 80 percent of American households read popular magazines in 1959. See Beth L. Bailey, *From Front Porch to Back Seat: Courtship in Twentieth-Century America* (Baltimore: Johns Hopkins University Press, 1988), 7.

14. For a discussion of "the revolt against medical authority," see Edward Shorter, *Bedside Manners: The Troubled History of Doctors and Patients* (New York: Simon and Schuster, 1985), 228–40. For a more thorough analysis of the challenges to physicians' cultural, political, and economic power in the 1970s and the resulting implications for the American health care system, see Starr, *Social Transformation*, 379–419.

Chapter One: Genesis of the Pill

1. Rosalind Rosenberg, *Divided Lives: American Women in the Twentieth Century* (New York: Hill and Wang, 1992), 147.

2. William H. Chafe, *The American Woman: Her Changing Social, Economic, and Political Roles, 1920–1970* (New York: Oxford University Press, 1972), 219.

3. "How to Be a Woman," *Seventeen* (date unknown), quoted in Brett Harvey, *The Fifties: A Women's Oral History* (New York: Harper Collins, 1993), 73.

4. Betty Friedan, *The Feminine Mystique* (New York: Dell, 1974), 11–27.

5. Women who wanted to use the birth control pill in the 1960s also had to ask their doctors for permission. See Chapter Two for a discussion of women's changing relationships with their physicians in the era of the pill.

6. A 1955 survey found that 70 percent of white married women aged eighteen to thirty-nine had used some form of contraception at one time or another; a similar study five years later found that number had risen to 81 percent. See Norman B. Ryder and Charles F. Westoff, *Reproduction in the United States, 1965* (Princeton: Princeton University Press, 1971), 107. For more detail, see also Ronald Freedman, Pascal K. Whelpton, and Arthur A. Campbell, *Family Planning, Sterility, and Population Growth* (New York: McGraw-Hill, 1959) and Pascal K. Whelpton, Arthur A. Campbell, and John E. Patterson, *Fertility and Family Planning in the United States* (Princeton: Princeton University Press, 1966).

7. Harvey, *The Fifties*, 11–12.

8. C. Thomas Dienes, *Law, Politics, and Birth Control* (Urbana: University of Illinois Press, 1972), 317–19.

9. *Griswold v. Connecticut;* see ibid., 162–83.

10. *Eisenstadt v. Baird;* see ibid., 245–52.

11. The statistics on physicians and contraception that follow are taken from Mary Jean Cornish, Florence A. Ruderman, and Sydney S. Spivak, *Doctors and Family Planning* (New York: National Committee on Maternal Health, 1963), 66–67, 31, 40, 19.

12. For a thorough study of the life and work of Margaret Sanger, see Ellen Chesler, *Woman of Valor: Margaret Sanger and the Birth Control Movement in America* (New York: Simon & Schuster, 1992).

13. Ibid., 407. Sanger had entertained notions of an oral contraceptive for decades. In a 1939 letter to Clarence Gamble, she wrote: "I've got herbs from Fiji which are said to be used to prevent conception. I'm hoping this may prove to be the 'magic pill' I've been hoping for since 1912 when women used to say, 'Do tell me the secret—can't I get some of the medicine, too?' I'm talking to one of the Squibb men and if they can analyze it for us we may find out 'somefin.'" Quoted in Greer Williams, "Biography of Clarence Gamble," unpublished manuscript, Schlesinger Library, Radcliffe College, Cambridge, Mass.

14. Quoted in Chesler, *Women of Valor*, 146.

15. Thomas Malthus, *An Essay on the Principle of Population* (1798; London: Penguin Books, 1970).

16. Linda Gordon provides an excellent analysis of the relationship between eugenics and population control, as well as the relationship of these two movements with the birth control movement, in her book, *Woman's Body, Woman's Right* (New York: Penguin Books, 1990). My discussion is limited to a brief outline of the roots of population control; for more information, see Gordon, 269–86, 386–96.

17. James Reed, *The Birth Control Movement and American Society: From Private Vice to Public Virtue* (Princeton: Princeton University Press, 1984), 282–83. See also Gordon, *Woman's Body, Woman's Right*, 393.

18. Reed, *Birth Control Movement*, 285.

19. *Summary Report,* Conference on Population Problems (June 20–22, 1952), 2, Rockefeller Family Archives, record group 2, John D. Rockefeller III Papers, box 43, population envelope.

20. Ibid., 3.

21. Elaine Moss, *The Population Council: A Chronicle of the First Twenty-Five Years, 1952–1977* (New York: Population Council, 1977), 14–15.

22. *Minutes of the Annual Meeting of the Members of the Corporation and of the Board of Trustees* (May 16, 1958), 1, Rockefeller Family Archives, record group 2, John D. Rockefeller III Papers, box 71.

23. Gordon, *Woman's Body, Woman's Right,* 386.

24. For an analysis of Planned Parenthood as a reform organization firmly entrenched within the contemporary social structure, see ibid., 337–40.

25. The role of Katherine McCormick in financing the pill project is discussed in greater detail later in this chapter.

26. Gregory Pincus, *The Control of Fertility* (New York: Academic Press, 1965), 194.

27. An earlier version of the intrauterine contraceptive device, the Grafenberg ring, had been available in the 1920s, but this model often caused pelvic infection, a serious condition in the preantibiotic era, so it was considered much too dangerous to use. The possibility of a modern IUD was not revisited until the late 1950s, and IUDs reentered the American market in the early 1960s.

28. Pincus's professional experience, which also contributed to his being a good person for the job, has been adequately treated elsewhere. For a thorough description and analysis of Pincus's career prior to the pill project, see Reed, *Birth Control Movement,* 317–33. For another analysis of Pincus's professional and personal motivations, see Stacey M. Berg, "Gregory Goodwin Pincus: From Reproductive Biology to the Birth Control Pill: Underlying Motivations and Professional Goals," (undergraduate honors thesis, Harvard University, 1989), 21–49.

29. Judith McGaw, "Review: Women and the History of American Technology," *Signs* 7 (Summer 1982): 803.

30. Adele Clarke, "Embryology and the Rise of American Reproductive Sciences," in *The Expansion of American Biology,* eds. Keith R. Benson, Jane Maienschein, and Ronald Rainger (New Brunswick, N.J.: Rutgers University Press, 1991), 107.

31. Ibid., 108. See also Adele Clarke, "Controversy and the Development of Reproductive Sciences," *Social Problems* 37 (February 1990): 20.

32. Merriley Borell, "Biologists and the Promotion of Birth Control Research, 1918–1938," *Journal of the History of Biology* 20 (Spring 1987): 53.

33. Clarke, "Embryology and the Rise of American Reproductive Sciences," 116.

34. Reed, *Birth Control Movement,* 313.

35. Clarke, "Embryology and the Rise of American Reproductive Sciences," 122, 110.

36. Borell, "Biologists and the Promotion of Birth Control Research," 53.

37. Frank B. Colton, "Steroids and 'the Pill': Early Steroid Research at Searle," *Steroids* 57 (December 1992): 625.

38. Carl Djerassi, *The Politics of Contraception* (New York: W. W. Norton, 1979), 239.

39. Colton, "Steroids and 'the Pill,'" 626; Carl Djerassi, "Steroid Research at Syntex: 'the Pill' and Cortisone," *Steroids* 57 (December 1992): 634.

40. For the information in this section, I am indebted to Reed, *The Birth Control Movement and American Society,* especially pp. 331–33, and 357–62.

41. Irwin C. Winter to Gregory Pincus (December 29, 1958). Gregory Pincus Papers, Library of Congress Manuscript Division (hereafter GPP-LC).

42. Djerassi, "Steroid research at Syntex," 636; Djerassi, *Politics of Contraception*, 251–52.

43. Worcester Foundation for Experimental Biology, annual reports. By 1956, the Foundation's annual income exceeded $1,000,000; the next year it increased to more than $1,600,000.

44. Clarke, "Controversy and the Development of Reproductive Sciences," 27. Clarke reports that NIH was not allowed to sponsor contraceptive research until 1959.

45. Gregory Pincus, correspondence with Planned Parenthood (1952, 1953), GPP-LC.

46. Gregory Pincus to Frederick Osborn (December 21, 1953); Frederick Osborn to Min-Chueh Chang (October 15, 1954), Population Council Records, accession 1, box 16, folder 280.

47. The following account is a summary of a more lengthy treatment in Reed, *Birth Control Movement*, 334–45.

48. Katherine McCormick to Margaret Sanger (January 22, 1952), quoted in ibid., 339.

49. William Vogt to Gregory Pincus (April 18, 1952), professional correspondence—PPF, 1952, Planned Parenthood Federation of America Records, Sophia Smith Collection, Northampton, Mass. (hereafter PPFA-SSC).

50. Gregory Pincus, *The Control of Fertility* (New York: Academic Press, 1965), dedication page.

51. J. Rock, G. Pincus, and C. R. Garcia, "Effects of Certain 19-nor Steroids on the Normal Human Menstrual Cycle," *Science* 124 (November 2, 1956), 891–93.

52. M. C. Chang, "Development of the Oral Contraceptives," *American Journal of Obstetrics and Gynecology* 132 (1978): 217.

53. Gregory Pincus and M. C. Chang, "The Effects of Progesterone and Related Compounds on Ovulation and Early Development in the Rabbit," *Acta Physiologica Latinoamericana* 3 (1953): 177–83.

54. Robert F. Slechta, M. C. Chang, and G. Pincus, "Effects of Progesterone and Related Compounds on Mating and Pregnancy in the Rat," *Fertility and Sterility* 5 (May–June 1954): 282.

55. Gregory Pincus, "Some Effects of Progesterone and Related Compounds upon Reproduction and Early Development in Mammals," *Report of the Proceedings of the Fifth International Conference on Planned Parenthood* (October 24–29, 1955, Tokyo, Japan), 175–184; Pincus et al., "Effects of Certain 19-Nor Steroids," 890–91; Gregory Pincus et al., "Studies of the Biological Activity of Certain 19-Nor Steroids in Female Animals," *Endocrinology* 59 (December 1956): 695–707.

56. Reed, *Birth Control Movement*, 351.

57. Quoted in ibid., 352.

58. John Rock, Celso Ramon Garcia, and Gregory Pincus, "Synthetic Progestins in the Normal Human Menstrual Cycle," *Recent Progress in Hormone Research* 13 (1957): 324.

59. Ibid., 324–25; Gregory Pincus, "Some Effects of Progesterone and Related Compounds upon Reproduction and Early Development in Mammals," *Acta Endocrinologica* Suppl. 28 (1956): 21–24.

60. Edward Tyler also conducted clinical trials of Syntex's norethindrone in Los Angeles, but the narrative here focuses on the work leading to the development of Searle's Enovid as the first birth control pill.

61. Rock, Garcia, and Pincus, "Synthetic Progestins," 323, 338–39.

62. Annette B. Ramirez and Conrad Seipp, *Colonialism, Catholicism, and Contraception* (Chapel Hill: University of North Carolina Press, 1983), 110.

63. David B. Tyler to Gregory Pincus (July 8, 1955), GPP-LC.

64. Gregory Pincus, "Long Term Administration of Enovid to Human Subjects," *Proceedings of a Symposium on 19-Nor Progestational Steroids* (Chicago: Searle Research Laboratories, 1957).

65. Reed, *Birth Control Movement*, 359.

66. Barbara Seaman, *The Doctors' Case against the Pill* (New York: Peter H. Wyden, 1969), 236–38; Gena Corea, *The Hidden Malpractice: How American Medicine Mistreats Women* (New York: Harper Colophon Books, 1985), 151; Gordon, *Woman's Body, Woman's Right*, 421.

67. Ramirez and Seipp, *Colonialism, Catholicism, and Contraception*, 108–23; Reed, *Birth Control Movement*, 359–62; Paul Vaughan, *The Pill on Trial* (New York: Coward-McCann, 1970), 38–53; McLaughlin, *The Pill, John Rock, and the Church*, 128–33; Edris Rice-Wray, "Field Study with Enovid as a Contraceptive Agent," *Proceedings of a Symposium on 19-Nor Progestational Steroids* (Chicago: Searle Research Laboratories, 1957), 78–85; Gregory Pincus, John Rock, and Celso R. Garcia, "Field Trials with Norethynodrel as an Oral Contraceptive," *Proceedings of the 6th International Conference on Planned Parenthood* (New Delhi, 1959), 216–30.

68. "Pills Are Distributed Against the Birth Rate," *El Imparcial* (April 21, 1956). GPP-LC.

69. Pincus, Rock, and Garcia, "Field Trials with Norethynodrel," 216.

70. Rice-Wray, "Field Study with Enovid," 85.

71. Edris Rice-Wray, "Twenty Years of Oral Contraception," *IPPF Medical Bulletin* 15 (February 1981): 1.

72. Reed, *Birth Control Movement*, 364.

73. The Kefauver Amendments strengthened the FDA's regulation of the introduction of drugs onto the market by giving the agency greater authority and control over the testing of new drugs. See Peter Temin, *Taking Your Medicine: Drug Regulation in the United States* (Cambridge: Harvard University Press, 1980), 120–26.

74. In June 1957, with no fanfare, the U.S. Food and Drug Administration approved Searle's application to market its formulation of norethynodrel with 1.5 percent mestranol (tradename Enovid) for the treatment of dysmenorrhea (painful menstruation), hypermenorrhea (excessive menstruation), and endometriosis (the growth of uterine tissue outside the uterus), based on clinical experience with about 600 gynecological patients. Other manufacturers (e.g., Parke-Davis) also marketed progestins as therapeutic agents for gynecological disorders, infertility, and habitual abortion.

75. Peter Montague, "New Perspectives on Toxics—Part 1: America Learns About Teratogens," *Rachel's Hazardous Waste News*, January 27, 1993, 1–2.

Chapter Two: Physicians, Patients, and the New Oral Contraceptives

1. Norman B. Ryder and Charles F. Westoff, *Reproduction in the United States, 1965* (Princeton: Princeton University Press, 1971), 141–142.

2. Norman B. Ryder and Charles F. Westoff, "Use of Oral Contraception in the United States, 1965," *Science* 153 (September 9, 1966): 1204.

3. "Annual Report Forms" (1961, 1963, 1964), annual reports folder, PPFA-SSC.

4. Nancy Van Vleck to Professional Staff (May 27, 1968), annual reports folder, PPFA-SSC.

5. James S. Coleman, Elihu Katz, and Herbert Menzel, *Medical Innovation: A Diffusion Study* (Indianapolis: Bobbs-Merrill, 1966), 59.

6. T. Caplow and J. J. Raymond, "Factors Influencing the Selection of Pharmaceutical Products," *Journal of Marketing* 19 (1954): 18–23, cited in Russell R. Miller, "Prescribing Habits of Physicians: A Review of Studies on Prescribing of Drugs," *Drug Intelligence and Clinical Pharmacy* 8 (February 1974): 84–85.

7. G. D. Searle & Company to Doctors (July 22, 1957), records of G. D. Searle & Company.

8. Alan Guttmacher and Carl G. Hartman, "Statement on Rock–Pincus Studies on Steroids" (July 10, 1957), PPWP Oral Contraceptive History—Nelson hearings background material, PPFA-SSC.

9. George J. Striker, "A Time for Reaping—Enovid 5 mg.," *The Searleman* (June 1961), reprinted in U.S. Senate Select Committee on Small Business (91st Cong., 2d sess.), *Hearings on Competitive Problems in the Drug Industry* (Washington, D.C.: Government Printing Office, 1970), 6268–69.

10. Striker, in *Hearings on Competitive Problems in the Drug Industry,* 6268–69.

11. The estimate of the number of detailmen is taken from Miller, "Prescribing Habits of Physicians," 82.

12. Miller, "Prescribing Habits of Physicians," 85.

13. *Enovid for Long-Term Ovulation Control* (Chicago: G. D. Searle & Company, 1960), Enovid—reprints, PPFA-SSC.

14. *Enovid for Ovulation Control* (Chicago: G. D. Searle & Company, 1963), records of G. D. Searle & Company.

15. *Obstetrics and Gynecology* 16 (November 1960): n.p.

16. Ibid. 18 (July 1961): 62–63.

17. Ibid.

18. Ibid. 19 (March 1962), 114–15.

19. Ibid. 20 (October 1962): n.p. The link between oral contraception and thromboembolism is treated more fully in Chapter Four.

20. *Journal of the American Medical Association* 181 (August 25, 1962): 172.

21. *Obstetrics and Gynecology* 21 (February 1963), n.p. For an interesting history and analysis of the Dialpak innovation, see Patricia Gossel, "Packaging the Pill," a paper presented at the Workshop on Museums and the History of Technology at the Science Museum of London (July 30, 1996).

22. *Obstetrics and Gynecology* 23 (April 1964): n.p.

23. Ibid., 22.

24. James Reed, *The Birth Control Movement and American Society* (Princeton: Princeton University Press, 1984), 364–65.

25. Edward T. Tyler, "Oral Contraception," *Journal of the American Medical Association* 175 (January 21, 1961): 225–26.

26. Mary S. Calderone, "Impact of New Methods on Practice in 73 Planned Parenthood Centers" (October 17, 1962), PPWP oral contraceptive history—Nelson hearings background material, PPFA-SSC.

27. Alan F. Guttmacher, "Statement on Enovid" (August 1960), Enovid—correspondence from 1957, PPFA-SSC.

28. Mary S. Calderone to J. William Crosson (May 25, 1960), Enovid—correspondence from 1957, PPFA-SSC.

29. Mary Calderone to all Planned Parenthood centers (January 23, 1961), Enovid—correspondence from 1961, PPFA-SSC.

30. Calderone, "Impact of New Methods of Practice in 73 Planned Parenthood Centers."

31. "Annual Report Forms" (1969), Annual reports folder, PPFA-SSC.

32. Ibid.

33. "Contraceptive Pill?" *Time*, May 6, 1957, 83.

34. Robert Sheehan, "The Birth-Control 'Pill'," *Fortune*, April 1958, 222.

35. Albert Q. Maisel, "Where Do We Stand with the Birth-Control Pill?" *Reader's Digest*, February 1961, 62.

36. "The 'Remarkable' Pill," *Newsweek*, January 30, 1961, 71.

37. "Birth Control Pills: The Full Story," *Good Housekeeping*, September 1962, 154.

38. John Rock, "It is Time to End the Birth-Control Fight," *Saturday Evening Post*, April 20, 1963, 14.

39. "The Pills," *Time*, February 17, 1961, 39.

40. Alan F. Guttmacher, "How Safe are Birth Control Pills?" *Ebony*, April 1962, 127.

41. The Ad Hoc Committee for the Evaluation of a Possible Etiologic Relation with Thromboembolic Conditions, "Final Report on Enovid." Submitted to the Commissioner of the Food and Drug Administration of the Department of Health, Education, and Welfare, September 12, 1963. Also published as "The Final Enovid Report," in *Journal of New Drugs* 3 (July–August 1963): 201–11.

42. Reproduced in *Time*, April 7, 1967, 84.

43. Worcester Foundation for Experimental Biology, *Annual Report* (1961–1962), 7–8. Reported in "Birth Control Pills: The Full Story," *Good Housekeeping*, September 1962, 155.

44. Robert K. Plumb, "Birth Pill Study Hits Stock Prices," *New York Times*, June 18, 1964, 37; William S. Fletcher, E. Douglas McSweeney Jr. and J. Englebert Dunphy, "The Effect of Various Hormones, Isotopes, and Isotope-Tagged Hormones on Induced Breast Cancer in the Rat," *Journal of the American Medical Association* 188 (May 1964): 430.

45. "Do the Pills Cause Cancer?" *Time*, July 3, 1964, 46.

46. "Enovid Exonerated," *Newsweek*, June 29, 1964, 80.

47. Ellen Willis, "The Birth-Control Pill," *Mademoiselle*, January 1961, 55.

48. Ibid., 113.

49. Gloria Steinem, "The Moral Disarmament of Betty Co-ed," *Esquire*, September 1962, 154, 155.

50. Ibid., 157.

51. Several lay Catholic periodicals (e.g., *America, Catholic World, Commonweal*) commented on the controversy over birth control. Sources for this discussion are restricted to nonsectarian newspapers and magazines because this analysis focuses on information directed to the general reading public.

52. Jack Star, "Catholics Take a New Look at the Pill," *Look*, September 8, 1964, 71.

53. William H. Shannon, *The Lively Debate: Response to Humanae Vitae* (New York: Sheed & Ward, 1970), 32–33.

54. John Rock, *The Time Has Come* (New York: Alfred A. Knopf, 1963), 168–69.

55. Robert E. Hall, "A Doctor's Answer to the Population Explosion," *New York Times Book Review,* May 12, 1963, 30.

56. John Rock, "Time to End the Birth-Control Fight," 10–14.

57. "Birth Control: The Pill and the Church," *Newsweek,* July 6, 1964, 52.

58. Jack Star, "Catholics Take a New Look at the Pill," 72.

59. "Birth Control: The Pill and the Church," *Newsweek,* July 6, 1964, 51.

60. *The Gallup Poll, Public Opinion 1935–1971* (New York: Random House, 1972), 1785–86, 1916, 1957.

61. See Charles F. Westoff and Larry Bumpass, "The Revolution in Birth Control Practices of U.S. Roman Catholics," *Science* 179 (January 5, 1973): 41–44. See also Bernard Asbell, *The Pill: A Biography of the Drug that Changed the World* (New York: Random House, 1995), 286–97.

62. "Birth Control Devices and Debates Engross the U.S." *Life,* May 10, 1963, 40.

63. "How Many Babies Is Too Many?" *Newsweek,* July 23, 1962, 27–34.

64. William L. Laurence, "Report on New Oral Contraceptive," *New York Times,* February 19, 1963, 6.

65. "Birth Control Pills Boom," *Business Week,* February 23, 1963, 62.

66. Ibid., 64.

67. "Birth Control Pills and Women in South," *New York Times,* November 11, 1964, 40.

68. "The Pills: More Effective, and More of Them," *Time,* March 20, 1964, 65.

69. J. D. Ratcliff, "An End to Woman's 'Bad Days'?" *Reader's Digest,* December 1962, 76.

70. Gregory Pincus, "Tell Me, Doctor," *Ladies' Home Journal,* June 1963, 134.

71. "More about Birth-Control Pills," *Good Housekeeping,* November 1962, 155.

72. "Birth Control Pills: The Full Story," *Good Housekeeping,* September 1962, 153.

73. Esther M. Shoemaker to Mary S. Calderone (March 27, 1961), Enovid—correspondence from 1961, PPFA-SSC.

74. Mrs. C. to Gregory Pincus (October 31, 1957), GPP-LC.

75. Mrs. R. to Gregory Pincus (no date), GPP-LC.

76. Mrs A. to Planned Parenthood (August 10, 1961), Enovid—individual requests for information, from 1960, PPFA-SSC.

77. Mr. and Mrs. H. to Gregory Pincus (October 27, 1960), GPP-LC.

78. Mrs. A. to George [sic] Pincus (June 28, 1957), GPP-LC.

79. Dr. B. to Dr. Paul Lavietes (n.d., probably May 1962), Rockefeller University, record group 891, *Medical Letter,* box 38, folder 7.

80. "Oral Contraceptives" *Medical Letter* (June 19, 1962), 2. Rockefeller University, record group 891, *Medical Letter,* box 38, folder 6.

81. William L. Searle to U.S. and Canadian Searlemen, Divisional Sales Managers, and Regional Sales Managers, August 9, 1962, reprinted in U.S. Senate, *Hearings on Competitive Problems in the Drug Industry,* 6273.

82. Quoted in Alice Lake, "The Pill," *McCall's,* November 1967, 96.

83. See Judith Walzer Leavitt, *Brought to Bed: Childbearing in America, 1750–1950* (New York: Oxford University Press, 1986).

84. Sheila M. Rothman, *Living in the Shadow of Death: Tuberculosis and the Social Experience of Illness in American History* (New York: Basic Books, 1994), 110.

85. Judith Walzer Leavitt, letter to author (October 12, 1994).

86. David J. Rothman, *Strangers at the Bedside: A History of How Law and Bioethics Transformed Medical Decision Making* (New York: Basic Books, 1991), 142; Paul Starr, *The Social Transformation of American Medicine* (New York: Basic Books, 1982), 380.

Chapter Three: Sex, Population, and the Pill

1. Linda Grant, *Sexing the Millennium: Women and the Sexual Revolution* (New York: Grove Press, 1994), 58.

2. Barbara Ehrenreich, Elizabeth Hess, and Gloria Jacobs, *Re-Making Love: The Feminization of Sex* (Garden City, N.Y.: Anchor Press/Doubleday, 1986), 41, 62.

3. John D'Emilio and Estelle B. Freedman, *Intimate Matters: A History of Sexuality in America* (New York: Harper & Row, 1988), xviii.

4. Ibid., 303. See also Ehrenreich, Hess, and Jacobs, *Re-Making Love*, 41.

5. The expression "the myth of the vaginal orgasm" was coined by Anne Koedt in a widely reprinted article of the same title that first appeared in 1969. In *Human Sexual Response,* Masters and Johnson concluded that "clitoral and vaginal orgasms are not separate entities," thus debunking the Freudian assumption that women who could achieve orgasms solely by vaginal stimulation were somehow more mature. See William H. Masters and Virginia C. Johnson, *Human Sexual Response* (Boston: Little, Brown, 1966), 67; and D'Emilio and Freedman, *Intimate Matters*, 312–13.

6. Blanche Linden-Ward and Carol Hurd Green, *American Women in the 1960s: Changing the Future* (New York: Twayne, 1993), xv.

7. "The Second Sexual Revolution," *Time* January 24, 1964, 54–59.

8. Drawing by Dedini, reproduced in *Newsweek* January 12, 1970, 66.

9. John F. Kantner and Melvin Zelnik, "United States: Exploratory Studies of Negro Family Formation—Common Conceptions about Birth Control," *Studies in Family Planning* No. 47 (November 1969), 13.

10. Dale McFeatters, "Women Rap Haden, Want Birth Control," *Pittsburgh Press,* August 7, 1968, 1, 10.

11. Alfred C. Kinsey, Wardell B. Pomeroy, and Clyde E. Martin, *Sexual Behavior in the Human Male* (Philadelphia: W. B. Saunders, 1948), 552.

12. Alfred C. Kinsey, Wardell B. Pomeroy, Clyde E. Martin, and Paul H. Gebhard, *Sexual Behavior in the Human Female* (Philadelphia: W. B. Saunders, 1953), 298.

13. Paul Robinson, *The Modernization of Sex* (New York: Harper Colophon Books, 1976), 103. See also Daniel Scott Smith, "The Dating of the American Sexual Revolution: Evidence and Interpretation," in *The American Family in Socio-Historical Perspective,* ed. Michael Gordon (New York: St. Martin's Press, 1973), 321–35.

14. Robert R. Bell and Jay B. Chaskes, "Premarital Sexual Experience among Coeds, 1958 and 1968," *Journal of Marriage and the Family* 32 (February 1970): 81–84; Harold T. Christensen and Christina F. Gregg, "Changing Sex Norms in America and Scandinavia," *Journal of Marriage and the Family* 32 (November 1970): 616–27.

15. Ira E. Robinson, Karl King, and Jack O. Balswick, "The Premarital Sexual Revolution Among College Females," *The Family Coordinator* 21 (April 1972): 189–94.

16. Ira Reiss, "Introduction," *Journal of Social Issues* 22 (April 1966): 3–4.

17. Vance Packard, *The Sexual Wilderness: The Contemporary Upheaval in Male–Female Relationships* (New York: David McKay, 1968), 17.

18. Ibid., 135–204.

19. Packard's study was the largest, involving 1393 Americans at 21 colleges and universities and 809 Europeans from four countries. See Packard, *Sexual Wilderness,* 152–60.

20. Christensen and Gregg, "Changing Sex Norms," 616.

21. Bell and Chaskes, "Premarital Sexual Experience," 81.

22. William Simon, "Review of *The Sexual Wilderness," American Sociological Review* 34 (August 1969): 605.

23. John H. Gagnon and William Simon, "Prospects for Change in American Sexual Patterns," *Medical Aspects of Human Sexuality* 4 (January 1970): 109.

24. John Gagnon, quoted in "Re-evaluating the Pill," *Newsweek,* January 12, 1970, 66.

25. Ira Reiss, *Premarital Sexual Standards* (New York: SIECUS, 1968), 6–7.

26. Bell and Chaskes, "Premarital Sexual Experience," 81.

27. Packard, *Sexual Wilderness,* 19–21.

28. Phillips Cutright, "The Teenage Sexual Revolution and the Myth of an Abstinent Past," *Family Planning Perspectives* 4 (January 1972): 29.

29. Charles F. Westoff and Norman B. Ryder, *The Contraceptive Revolution* (Princeton: Princeton University Press, 1977), 4.

30. Ronald Freedman, Pascal K. Whelpton, and Arthur A. Campbell, *Family Planning, Sterility, and Population Growth* (New York: McGraw-Hill, 1959). Pascal K. Whelpton, Arthur A. Campbell, and John E. Patterson, *Fertility and Family Planning in the United States* (Princeton: Princeton University Press, 1966).

31. Norman B. Ryder and Charles F. Westoff, *Reproduction in the United States, 1965* (Princeton: Princeton University Press, 1971), 107.

32. Whelpton, Campbell, and Patterson, *Fertility and Family Planning,* 285–87.

33. Ryder and Westoff, *Reproduction in the United States, 1965,* 122, 124.

34. Westoff and Ryder, *The Contraceptive Revolution,* 19.

35. The pill remained the most popular method of birth control throughout the 1970s, 1980s, and 1990s, despite the recurrent controversies over its adverse health effects.

36. G. D. Searle & Company to Doctors (January 17, 1963), Population Council Records, accession 1, box 90, folder 1696.

37. Harry L. Levin to Alan Guttmacher (July 22, 1966). Population Council—general correspondence from 1961, PPFA-SSC.

38. Advisory Committee on Obstetrics and Gynecology, Food and Drug Administration, *Second Report on the Oral Contraceptives* (Washington, DC: U.S. Government Printing Office, August 1, 1969), 13.

39. Charles F. Westoff and Larry Bumpass, "The Revolution in Birth Control Practices of U.S. Roman Catholics," *Science* 179 (January 5, 1973): 42.

40. Quoted in Ryder and Westoff, *Reproduction in the United States, 1965,* 217.

41. William D. Mosher and Charles F. Westoff, *Trends in Contraceptive Practice: United States, 1965–1976* (Hyattsville, Md.: National Center for Health Statistics, 1982), 15.

42. Ryder and Westoff, *Contraception in the United States, 1965,* 152–53. See also Ryder and Westoff, "The United States: The Pill and the Birth Rate, 1960–1965," *Studies in Family Planning* No. 20 (June 1967): 1–3.

43. Westoff and Ryder, *The Contraceptive Revolution,* 29.

44. Leslie Aldridge Westoff and Charles F. Westoff, *From Now to Zero* (Boston: Little, Brown, 1968), 220.

45. John F. Kantner and Melvin Zelnick, "Sexual Experience of Young Unmarried

Women in the United States," *Family Planning Perspectives* 4 (October 1972): 9.

46. John F. Kantner and Melvin Zelnick, "Contraception and Pregnancy: Experience of Young Unmarried Women in the United States," *Family Planning Perspectives* 5 (Winter 1973): 21–22.

47. Ibid., 22.

48. Ibid., 26.

49. Quoted in "The Pill on Campus," *Newsweek*, October 11, 1965, 93.

50. Andrew Hacker, "The Pill and Morality," *New York Times Magazine*, December 12, 1965, 140.

51. "The Pill: How It Is Affecting U.S. Morals, Family Life," *U.S. News & World Report*, July 11, 1966, 62.

52. Pearl S. Buck, "The Pill and the Teen-Age Girl," *Reader's Digest*, April 1968, 111.

53. Frederic R. Talbot, "Should Doctors Prescribe Contraceptives for Unmarried Girls?" *Ladies' Home Journal*, January 1968, 87.

54. "Freedom from Fear," *Time*, April 7, 1967, 80.

55. Hacker, "The Pill and Morality," 140.

56. Leona Baumgartner, "Government Responsibility for Family Planning in the United States," in *Fertility and Family Planning: A World View*, eds. S. J. Behrman, Leslie Corsa, Jr., and Ronald Freedman (Ann Arbor: University of Michigan Press, 1969), 440.

57. Fred Jaffe to Charles Westoff, Arthur Campbell, and Ronald Freedman (December 2, 1965), Population Council Records, accession 1, box 107, folder 1907.

58. Jeannie Rosoff to PP-WP Board Members, Affiliate Presidents and Executive Directors, Clergymen's, Medical, and Social Science Advisory Committees (November 2, 1966), Population Council Records, accession 1, box 116, folder 2133.

59. U.S. President's Committee to Study the U.S. Military Assistance Program, *Composite Report* 1 (Washington, D.C.: August 17, 1959), 94–97.

60. *New York Times*, December 3, 1959, 1, 18.

61. *The Gallup Poll, Public Opinion 1935–1971* (New York: Random House, 1972), 1654.

62. Nancy Aries, "Fragmentation and Reproductive Freedom: Federally Subsidized Family Planning Services, 1960–1980," *American Journal of Public Health* 77 (November 1987): 1465.

63. Phyllis Tilson Piotrow, *World Population Crisis: The United States Response* (New York: Praeger, 1973), 46, 88.

64. "Statements by President Johnson Concerning Population," Population Council Records, accession 1, box 115, folder 2109.

65. Raymond A. Lamontagne to John D. Rockefeller III (September 22, 1965), Population Council Records, accession 1, box 116, folder 2131.

66. Aries, "Fragmentation and Reproductive Freedom," 1466–67.

67. Senator Ernest Gruening to Bernard Berelson and Frank Notestein (December 22, 1966), Population Council Records, accession 1, box 115, folder 2109.

68. "Summary of federal financial participation in birth control and/or family planning activities" (November 2, 1966), Population Council Records, accession 1, box 115, folder 2109.

69. Phyllis Tilson Piotrow, *World Population Crisis: The United States Response* (New York: Praeger Publishers, 1973). See also Frederick S. Jaffe, "Public Policy on Fertility Control," *Scientific American* 229 (July 1973): 17–23; and Oscar Harkavy, *Curbing Population*

Growth: An Insider's Perspective on the Population Movement (New York: Plenum Press, 1995).

70. Wylda B. Cowles, "Report to the Medical Committee, June, 1963, to May, 1964," Population Council Records, accession 1, box 107, folder 1995.

71. Alexander D. Langmuir to Frank Notestein (July 2, 1964), Population Council Records, accession 1, box 116, folder 2118.

72. Leslie Corsa, "United States: Public Policy and Programs in Family Planning," *Studies in Family Planning* No. 27 (March 1968), 4.

73. Margaret Snyder to Fred Jaffe (April 18, 1962), Population Council Records, accession 1, box 87, folder 1629.

74. "News of the Pill," *Time*, April 21, 1967, 66.

75. "Grant Files," Population Council Records, accession 1, boxes 50–73.

76. Alan F. Guttmacher to John G. Searle (December 29, 1964), PPFA: G. D. Searle & Company, correspondence from 1964, PPFA-SSC.

77. Alan F. Guttmacher, "The Pill or the IUD?" *President's Letter* (January 26, 1965), 3, Alan F. Guttmacher Papers, Francis A. Countway Library of Medicine (hereafter AGP-CLM).

78. Westoff and Ryder, *Contraceptive Revolution*, 34, 41. Even at the height of its popularity in the 1970s, the IUD was never used by more than 10 percent of American women.

Chapter Four: Debating the Safety of the Pill

1. Joyce Avrech Berkman, "Historical Styles of Contraceptive Advocacy," in *Birth Control and Controlling Birth: Women-Centered Perspectives*, eds. Helen B. Holmes, Betty B. Hoskins, and Michael Gross (Clifton, N.J.: Humana Press, 1980), 27–28.

2. Paul Starr, *The Social Transformation of American Medicine* (New York: Basic Books, 1982), 379. See also David Rothman, *Strangers at the Bedside* (New York: Basic Books, 1991), 141–44; John Duffy, *From Humors to Medical Science: A History of American Medicine* (Chicago: University of Illinois Press, 1993), 323–24; and Edward Shorter, *Bedside Manners: The Troubled History of Doctors and Patients* (New York: Simon and Schuster, 1985), 228–29.

3. Allan Mazur, *The Dynamics of Technical Controversy* (Washington, D.C.: Communications Press, 1981), 7–8.

4. I disagree with Mazur's interpretation of the public controversy over the safety of oral contraceptives. He describes birth control pills, along with color televisions, microwave ovens, medical and dental X-rays, food additives, hybrid seeds, and computer data banks, as technologies which, in spite of warnings issued, did not capture the attention of the American public. See Mazur, *Dynamics of Technical Controversy*, 62, 85, 90.

5. Richard D. Lyons, "New Data on the Safety of the Pill," *New York Times*, September 7, 1969, Sec. IV, 16.

6. Gregory Pincus, John Rock, and Celso R. Garcia, "Field Trials with Norethynodrel as an Oral Contraceptive," *Proceedings of the Sixth International Conference on Planned Parenthood* (New Delhi, 1959), 221–22.

7. From 1960 to 1965, 32 percent of pill users stopped within one year and 47 percent stopped within two years. Of this group of "dropouts," 65 percent discontinued oral contraception because of "unpleasant or frightening" side effects. Leslie Aldridge Westoff and

Charles F. Westoff, *From Now to Zero: Fertility, Contraception and Abortion in America* (Boston: Little, Brown, 1968), 109–10.

8. See Scarlett Pollock, "Refusing to Take Women Seriously: Side Effects and the Politics of Contraception," in *Test-Tube Women: What Future for Motherhood?* eds. Rita Arditti, Renate Duelli Klein, and Shelley Minden (London: Pandora Press, 1984), 148–51.

9. Roy Hertz, "Experimental and Clinical Aspects of the Carcinogenic Potential of Steroid Contraceptives," *International Journal of Fertility* 13 (October–December 1968): 274.

10. Daniel Seigel and Philip Corfman, "Epidemiological Problems Associated with Studies of the Safety of Oral Contraceptives," *Journal of the American Medical Association* 203 (March 1968): 148–52.

11. For a critique of the notion of pregnancy as a disease, see Donald Merkin, *Pregnancy as a Disease: The Pill in Society* (Port Washington, N.Y.: Kennikat Press, 1976).

12. Morton Mintz argued convincingly that the pill's safety was relative to the individual's medical situation and her ability to use alternative methods of birth control effectively. See, for example, Morton Mintz, "The Golden Pill: We Can't Yet Be Sure It's Safe," *New Republic*, March 2, 1968, 18–20.

13. Even in the 1990s, experts continued to debate the relationship between oral contraceptives and cancer. See David A. Grimes, "The Safety of Oral Contraceptives: Epidemiological Insights from the First 30 Years," *American Journal of Obstetrics and Gynecology* 166 (June 1992): 1950–54.

14. Barbara Seaman and Gideon Seaman, *Women and the Crisis in Sex Hormones* (New York: Bantam Books, 1977), 101.

15. G. D. Searle & Company, *Proceedings of a Conference: Thromboembolic Phenomena in Women* (Chicago: 1962).

16. John Rock, quoted in ibid., 114.

17. Irwin C. Winter, quoted in ibid., 117.

18. G. D. Searle & Company to Doctors (December 26, 1962), Enovid—correspondence from July 1962, PPFA-SSC.

19. The Ad Hoc Committee for the Evaluation of a Possible Etiologic Relation with Thromboembolic Conditions, "Final Report on Enovid," submitted to the Commissioner of the Food and Drug Administration of the Department of Health, Education, and Welfare, September 12, 1963, 2. Also published as "The Final Enovid Report," in *Journal of New Drugs* 3 (July–August 1963): 201–11.

20. "Wright Committee Reports on Enovid," U.S. Senate Select Committee on Small Business, Subcommittee on Monopoly, *Competitive Problems in the Drug Industry* (Washington, D.C.: Government Printing Office, 1970), 7240.

21. "Final Report on Enovid," 14. See also "Researchers Call FDA Warning on Enovid Wrong, Blame Own Mathematical Error," *Wall Street Journal*, September 18, 1963, 28.

22. Worcester Foundation for Experimental Biology, *Annual Report* (1961–1962), 7–8.

23. Gregory Pincus and Celso-Ramon Garcia, "Studies on Vaginal, Cervical and Uterine Histology," *Metabolism* 14 (March 1965): 346.

24. Advisory Committee on Obstetrics and Gynecology, Food and Drug Administration, *Report on the Oral Contraceptives* (Washington, D.C.: Government Printing Office, August 1, 1966), 21 (hereafter *Report on the Oral Contraceptives*).

25. William S. Fletcher, E. Douglas McSweeney, Jr., and J. Englebert Dunphy, "The

Effect of Various Hormones, Isotopes, and Isotope-Tagged Hormones on Induced Breast Cancer in the Rat," *Journal of the American Medical Association* 188 (May 1964): 430.

26. E. Douglas McSweeney and William S. Fletcher, "Synthetic Estrogen–Progestin Combinations: Effect on Hormone-sensitive Breast Cancer in the Rat," *Archives of Surgery* 99 (November 1969): 652–54.

27. "Oral Contraceptives," *Journal of the American Medical Association* 195 (February 1966): 31.

28. Frank B. Walsh et al., "Oral Contraceptives and Neuro-Ophthalmologic Interest," *Archives of Ophthalmology* 74 (November 1965): 628.

29. Ben A. Franklin, "Birth-pill makers agree to give warning on possible eye risks," *New York Times*, November 18, 1965, 1.

30. Steven M. Spencer, "The Birth Control Revolution," *Saturday Evening Post*, January 15, 1966, 24.

31. Jane E. Brody, "The Pill: Revolution in Birth Control," *New York Times*, May 31, 1966, 34.

32. Lawrence Lader, "Three Men Who Made a Revolution," *New York Times Magazine*, April 10, 1966, 58.

33. Ibid.

34. John Devaney, "How Safe are the Birth Control Pills?" *Redbook*, February 1963, 140.

35. Morton Mintz, interview by author, tape recording, Chevy Chase, Md., December 10, 1993.

36. Barbara Seaman, interview by author, tape recording, New York, N.Y., January 30, 1994.

37. Seaman, interview, "The Pros and Cons of the Pill," *Time*, May 2, 1969, 58.

38. "Clinical Aspects of Oral Gestogens," *World Health Organization Technical Report Series No. 326* (Geneva: WHO, 1966), 22.

39. Ibid., 3.

40. *Report on the Oral Contraceptives*, 13.

41. Roy Hertz, "An appraisal of certain problems involved in the use of steroid compounds for contraception," *Report on the Oral Contraceptives*, 49–59; Roy Hertz and John C. Bailar III, "Estrogen–Progestogen Combinations for Contraception," *Journal of the American Medical Association* 198 (November 1966): 136–42; Roy Hertz, "Experimental and Clinical Aspects of the Carcinogenic Potential of Steroid Contraceptives," *International Journal of Fertility* 13 (October–December, 1968): 273–86.

42. Hertz, "An appraisal of certain problems," 51.

43. AMA Council on Drugs, "Evaluation of Oral Contraceptives," *Journal of the American Medical Association* 199 (February 1967): 144–47.

44. According to the AMA's Annual Report of 1969, the number of physicians in the United States was estimated to be 327,000 and the number of AMA members was estimated to be 218,000. "The Search—AMA Annual Report, 1969," *Journal of the American Medical Association* 211 (January 1970): n.p.

45. AMA Council on Drugs, "Evaluation of Oral Contraceptives," 144.

46. "The Safe and Effective Pills," *Time*, August 19, 1966: 59; "Popular, Effective, Safe," *Newsweek*, August 22, 1966: 92; "Giving Pill a Safe Label," *Business Week*, August 13, 1966: 33.

47. "Health Report on 'The Pill'," *New York Times*, August 18, 1966, 34.

48. "Freedom from Fear," *Time* 89, April 7, 1967, 78, 79.

49. Lord Platt et al., "Risk of Thromboembolic Disease in Women Taking Oral Contraceptives," *British Medical Journal* 2 (May 1967): 355–59; W. H. W. Inman and M. P. Vessey, "Investigation of Deaths from Pulmonary, Coronary, and Cerebral Thrombosis and Embolism in Women of Child-bearing Age," *British Medical Journal* 2 (April 1968): 193–99; M. P. Vessey and Richard Doll, "Investigation of Relation Between Use of Oral Contraceptives and Thromboembolic Disease," *British Medical Journal* 2 (April 1968): 199–205.

50. Victor A. Drill and David W. Calhoun, "Oral Contraceptives and Thromboembolic Disease," *Journal of the American Medical Association* 206 (September 1968): 77–84.

51. Ibid., 83; Richard Doll, W. H. W. Inman, and M. P. Vessey, "Concerning the British Data," *Journal of the American Medical Association* 207 (February 1969): 1150. See also Carl C. Seltzer, "An Editorial Viewpoint," *Journal of the American Medical Association* 207 (February 1969): 1152.

52. Louis E. Moses, "Oral Contraceptives and Thromboembolism," *Journal of the American Medical Association* 208 (April 1969): 694.

53. Cecil Hougie, "Thromboembolic Disorders and Oral Contraceptives," *Journal of the American Medical Association* 208 (May 1969): 865.

54. Philip E. Sartwell, Alfonse T. Masi, Federico G. Arthes, Gerald R. Greene, and Helen E. Smith, "Thromboembolism and Oral Contraceptives: An Epidemiological Case-Control Study," in Advisory Committee on Obstetrics and Gynecology, Food and Drug Administration, *Second Report on the Oral Contraceptives* (Washington, D.C.: Government Printing Office, August 1, 1969), Appendix 2A (hereafter *Second Report on the Oral Contraceptives*). This paper was also published in the *American Journal of Epidemiology* 90 (November 1969): 365–80.

55. Sartwell et al., "Thromboembolism and Oral Contraceptives," *Second Report on the Oral Contraceptives*, 23.

56. "The Search for a Birth Control Method to Replace the Pill," *Good Housekeeping*, September 1967, 179–81; "The Pill and Strokes," *Time*, December 29, 1967, 32; "Perils of the Pill," *Newsweek*, May 13, 1968, 66; "Doubts about the Pill," *Newsweek*, May 19, 1969, 118.

57. "The Pill and Strokes," *Time*, December 29, 1967, 33.

58. Jane Brody, "Birth Control Pills: A Balance Sheet on their National Impact," *New York Times*, March 23, 1969, 60.

59. Lois R. Chevalier and Leonard Cohen, "The Terrible Trouble with the Birth-Control Pills," *Ladies' Home Journal*, July 1967, 43.

60. Morton Mintz, "The Golden Pill; We Can't Yet Be Sure It's Safe," *New Republic*, March 2, 1968, 20.

61. Brody, "Birth Control Pills," 60.

62. "How Safe Is 'the Pill'?" *New York Times*, September 12, 1969, 42.

63. Myron R. Melamed et al., "Prevalence Rates of Uterine Cervical Carcinoma in situ for Women Using the Diaphragm or Contraceptive Oral Steroids," *British Medical Journal* 3 (July 1969): 195–200. This study of the *prevalence* of cervical cancer examined the proportions of women who were diagnosed with a precancerous condition at the beginning of the study and thus differed from a *prospective* study which would determine the incidence of cancer as women under observation developed the disease over the course of many years or a *retrospective* study, which would compare the proportions using oral contraceptives in matched groups of women with cervical cancer and women without cervical cancer.

64. George Langmyhr to Alan Guttmacher and Paul Todd (December 10, 1968), medical department—Melamed–Dubrow affair 1968–69, PPFA-SSC.

65. Robert V. P. Hutter to Seymour L. Romney (February 12, 1969), medical department—Melamed–Dubrow affair 1968–69, PPFA-SSC.

66. George Langmyhr to Lynn Landman, Dr. Guttmacher, and Paul Todd (May 16, 1969), medical department—Melamed–Dubrow affair 1968–69, PPFA-SSC.

67. "The Pill and Cancer," *Newsweek*, August 11, 1969, 59; Robert V. P. Hutter to George J. Langmyhr (July 31, 1969), medical department—Melamed–Dubrow affair 1968–69, PPFA-SSC.

68. Hilton A. Salhalnick, David M. Kipnis, and Raymond L. Vande Wiele, eds., *Metabolic Effects of Gonadal Hormones and Contraceptive Steroids* (New York: Plenum Press, 1969).

69. Ibid., vii–ix.

70. *Second Report on the Oral Contraceptives*, 7.

71. *Report on the Oral Contraceptives*, 13.

72. *Second Report on the Oral Contraceptives*, 1.

73. Ibid., 1–2.

74. "A Report on the Questionnaire 'Survey of Experience with Oral Contraceptive Pills'" (October 1967), 1–6, pill survey folder, Records of the American College of Obstetricians and Gynecologists.

75. "What the Nation's Obstetricians Think about the Pill," *Ladies' Home Journal*, July 1967, 45.

76. "A Report on the Questionnaire," 4–5.

77. "What the Nation's Obstetricians Think about the Pill," 45.

78. "Control of Fertility," *Journal of the American Medical Association* 199 (February 1967): 155.

79. "The Pill's Grim (?) Progress," *Journal of the American Medical Association* 206 (September 1968): 124–25.

80. Nea D'Amelio, "How 334 Gynecologists View the Pill," *Medical Times* 98 (October 1970): 206–12.

81. "Our Readers Talk Back about the . . . Birth Control Pills," *Ladies' Home Journal*, November 1967, 161.

82. Ibid, 92.

83. Ibid.

84. Dorothy L. Millstone to James Irwin (May 9, 1969), PPFA: G. D. Searle & Company, correspondence from 1966, PPFA-SSC.

85. Richard L. Day and George Langmyhr, "Statement" (June 30, 1967), 1. PPFA: contraceptives, oral—miscellaneous material from 1960, PPFA-SSC.

86. Winfield Best and George Langmyhr to Affiliate Executives and PPWP Committees (November 29, 1967), PPWP oral contraceptive history—Nelson hearings background material, PPFA-SSC.

87. Dorothy L. Millstone to James Irwin (May 9, 1969), PPFA: G. D. Searle & Company, correspondence from 1966, PPFA-SSC.

88. Charles F. Westoff and Norman B. Ryder, *The Contraceptive Revolution* (Princeton: Princeton University Press, 1977), 44–48. Women who discontinued the pill to get pregnant or because they no longer needed contraception were excluded from this analysis.

89. *The Gallup Poll, Public Opinion 1935–1971* (New York: Random House, 1972), 2044–45, 2239.

90. Quoted in Barbara Seaman, *The Doctors' Case against the Pill* (New York: Peter H. Wyden, 1969), 54–55.

91. Quoted in Seaman, *Doctors' Case against the Pill,* 15.

92. Chevalier and Cohen, "Terrible Trouble with the Birth-Control Pills," 50.

93. Quoted in Alice Lake, "The Pill," *McCall's,* November 1967, 169.

Chapter Five: Oral Contraceptives and Informed Consent

1. Dorothy L. Millstone to Drs. Guttmacher, Langmyhr and Rogers, and Jack Scanlan (September 18, 1969); Dorothy L. Millstone to Dr. George Langmyhr (September 24, 1969), contraceptives, oral—general correspondence from 1965, PPFA-SSC.

2. Medical Department and Information & Education Department to Affiliates Executive Directors, Medical Directors, Presidents and the Board (October 22, 1969), PP-WP: Oral contraceptive history—Nelson hearings background material, PPFA-SSC.

3. Robert N. Mayer, *The Consumer Movement: Guardians of the Marketplace* (Boston: Twayne, 1989), 12–30.

4. A recent history traces the origins of the idea of informed consent in American medicine to the late 1950s and early 1960s. See Ruth R. Faden and Tom L. Beauchamp, *A History and Theory of Informed Consent* (New York: Oxford University Press, 1986), 86.

5. Boston Women's Health Book Collective, *Our Bodies, Ourselves* (Boston: New England Free Press, 1971), 1, quoted in Sheila M. Rothman, *Woman's Proper Place: A History of Changing Ideals and Practices, 1870 to the Present* (New York: Basic Books, 1978), 282.

6. In 1960, the Food and Drug Administration had approved the sale of Enovid for contraceptive purposes quietly and without public controversy. The current discussion refers to government involvement in the debate over the safety of the pill at the end of the decade.

7. Charles Mann, "Women's Health Research Blossoms," *Science* 269 (August 11, 1995), 766–70; Leora Tanenbaum, "Pill Politics," *Boston Phoenix,* February 24, 1995, 16, 18–19.

8. *New York Times,* August 20, 1994, 24; *Congressional Record,* May 2, 1995, E913.

9. Barbara Seaman, interviewed by the author, tape recording, New York, N.Y., January 30, 1994.

10. Christopher Lehmann-Haupt, "Pill Talk," *New York Times,* January 2, 1970, 27.

11. Seaman, interviewed by the author, New York, N.Y., January 30, 1994.

12. The subcommittee continued to hold hearings on "Competitive Problems in the Drug Industry" during the 1970s after the well-publicized series on the oral contraceptives in 1970. The published record of this decade of hearings eventually filled more than thirty volumes.

13. U.S. Congress, Senate Subcommittee on Monopoly, *Competitive Problems in the Drug Industry* (Washington, D.C.: U.S. Government Printing Office, 1967), 2.

14. Milton Silverman and Philip R. Lee, *Pills, Profits, and Politics* (Berkeley: University of California Press, 1974), 145.

15. *Competitive Problems in the Drug Industry,* 2753.

16. Seaman, interviewed by the author, New York, N.Y., January 30, 1994.

17. *Competitive Problems in the Drug Industry,* 5923.

18. Ibid., 6818.

19. Ben Gordon, interview by author, tape recording, Washington, D.C., February 17, 1994.

20. Ibid.

21. Seaman, interviewed by the author, New York, N.Y., January 30, 1994.

22. Female scientists did testify during the second set of hearings in February and March. Alice Wolfson, telephone interview by author, tape recording, October 25, 1994.

23. *Competitive Problems in the Drug Industry,* 5924.

24. Ibid., 6151.

25. ABC evening news (January 14, 1970), Vanderbilt Television News Archive (hereafter VTNA).

26. CBS evening news (January 14, 1970), VTNA.

27. Ibid.

28. NBC Huntley–Brinkley Report (January 14, 1970), VTNA. Davis's testimony against the pill was in part motivated by his interest in the Dalkon Shield, an intrauterine device he had helped to develop. His financial stake in the IUD, which he denied at the time of the hearings, later became evident.

29. ABC evening news and CBS evening news (January 14, 1970), VTNA.

30. ABC evening news (January 15, 1970), VTNA.

31. ABC evening news, NBC Huntley–Brinkley Report, CBS evening news (January 23, 1970), VTNA.

32. ABC evening news (January 23, 1970), VTNA.

33. NBC Huntley–Brinkley Report (January 22, 1970), VTNA.

34. NBC Huntley–Brinkley Report (January 23, 1970), VTNA.

35. NBC Huntley–Brinkley Report (January 19, 1970), VTNA; "FDA Bids Doctors Tell of Risks in Birth Pill Use," *New York Times,* (January 20, 1970), 1.

36. NBC Huntley–Brinkley Report (January 15, 1970), VTNA.

37. Landman and Millstone to Guttmacher, Langmyhr, and Scanlan (December 23, 1969), contraceptives, oral—general correspondence from 1965, PPFA-SSC.

38. Ibid.

39. Medical Department to Regional Directors, Executive Directors, Medical Directors, Medical Advisory Committee Chairmen, Information and Education Directors, Presidents, and PPWP Board (January 8, 1970), contraceptives, oral, misc. material from 1960, PPFA-SSC.

40. Dr. Alan F. Guttmacher to Regional Directors, Executive Directors, Medical Directors, Medical Advisory Committee Chairmen, Information and Education Directors, Presidents, and PPWP Board (January 16, 1970), Guttmacher—Nelson hearings, mimeos—general, PPFA-SSC.

41. Medical Department and the Information and Education Department to Executive Directors, Regional Directors, Medical Directors, Presidents, PPWP Board, Chairmen Medical Advisory Committee, I&E Directors (January 30, 1970), contraceptives, oral—misc. material from 1960, PPFA-SSC.

42. Dr. George Langmyhr to Lynn Landman (January 21, 1970), PPWP—oral contraceptive history—Nelson hearings background material, PPFA-SSC.

43. "Distinguished Physician Maintains Birth Pill Protects Women's Lives, Averts Illegal Abortion, Unwanted Births," Guttmacher, Nelson hearings, press release, February 25, 1970, PPFA-SSC.

44. Alan F. Guttmacher, "The Pill," *Wellesley Alumnae Magazine,* February 10, 1970, manuscript, 7, AGP-CLM.

45. President's Letter to Friends of Planned Parenthood (February 14, 1970), AGP-CLM.

46. Lynn Landman to Alan Guttmacher (January 30, 1970), Nelson hearings source material, PPFA-SSC.

47. According to the *New York Times,* four members of D.C. Women's Liberation hissed Guttmacher during his testimony before the Senate subcommittee. "Expert Decries 'Alarm' on Birth-Curb Pill," *New York Times,* February 26, 1970, 50.

48. *Competitive Problems in the Drug Industry,* 6632.

49. Alan F. Guttmacher Papers, CLM.

50. Cartoon by Bassett, in *Washington Daily News,* January 21, 1970.

51. "Poll on the Pill," *Newsweek,* February 9, 1970, 52–53.

52. George Langmyhr to Alan Guttmacher (March 4, 1970), Guttmacher, impact of Nelson hearings, PPFA-SSC.

53. "Fear of the Pill Aids an Industry," *Business Week,* March 21, 1970, 89.

54. Jane E. Brody, "Pregnancies Follow Birth Pill Publicity," *New York Times,* February 15, 1970, 28.

55. Dr. Howard A. Rusk, "Layman's Peace of Mind," *New York Times,* February 22, 1970, 91.

56. The Patient's Bill of Rights, adopted by the American Hospital Association in 1972, was developed out of this growing concern for informed consent and patient participation in medicine. See Paul Starr, *The Social Transformation of American Medicine* (New York: Basic Books, 1982), 389–91.

57. Phyllis Piotrow, in *Competitive Problems in the Drug Industry,* 6647; Morton Mintz, "The Pill and the Public's Right to Know," *The Progressive,* May 1970, 25.

58. Cartoon by Joseph Farris, in *Time,* March 9, 1970, 32.

59. Dr. Elizabeth Connell, in *Competitive Problems in the Drug Industry,* 6518.

60. *Competitive Problems in the Drug Industry,* 6800.

61. The testimony of a fifth woman, Dr. Anna Southam, was read into the record without additional comment and inquiry because she was unable to appear before the committee at her scheduled time.

62. *Competitive Problems in the Drug Industry,* 6522.

63. Ibid., 6524.

64. Opening statement of Washington Women's Liberation at Women's Hearings on the Birth Control Pill, March 7, 1970, in *Competitive Problems in the Drug Industry,* 7284.

65. Ibid., 7283.

66. In 1968, the FDA required a two-sentence warning label to be placed on containers of an asthma inhalant, isoproterenol. Since the container could be refilled by the pharmacist without the patient revisiting the physician, a warning was considered necessary to remind the patient of the danger of overdosage.

67. "Notes on the Package Insert," *Journal of the American Medical Association* 207 (February 1969): 1335.

68. "The Package Insert," *Journal of the American Medical Association* 207 (February 1969): 1342.

69. Peter Temin, *Taking Your Medicine: Drug Regulation in the United States* (Cambridge: Harvard University Press, 1980), 46–47.

70. "Text of the Proposed Leaflet on Birth Control Pills," *New York Times,* March 5, 1970, 24.

71. "F.D.A. Restricting Warning on Pill," *New York Times,* March 24, 1970, 8.

72. *Federal Register* 35 (April 10, 1970): 5962.

73. Alice Wolfson, "more on the pill," *off our backs,* April 11, 1970, 6.

74. Seaman, interviewed by the author, New York, N.Y., January 30, 1994.

75. Wolfson, telephone interview by the author, October 25, 1994.

76. Wolfson, "more on the pill," 6.

77. LGH to Hearing Clerk (April 24, 1970), records of the U.S. Food and Drug Administration (hereafter records-FDA).

78. JK to Secretary Finch (April 7, 1970), records-FDA.

79. FMR to Hearing Clerk (n.d.), records-FDA.

80. GW to Hearing Clerk (n.d.), records-FDA.

81. The Population Council also kept quiet on the issue of the proposed patient package insert for oral contraceptives; it never became involved in the medical or public controversy over the safety of oral contraceptives. Instead, its efforts focused on biomedical research, demographic studies, and technical assistance to family planning programs in developing nations.

82. In 1968, 76 percent of the women receiving birth control at more than five hundred Planned Parenthood affiliate clinics used oral contraceptives. Nancy Van Vleck to Professional Staff (May 7, 1969), annual reports folder, PPFA-SSC.

83. "Final Warning?" *Newsweek,* June 22, 1970, 76.

84. Dr. E.S.R. to Secretary Robert H. Finch (April 25, 1970), records-FDA.

85. Dr. S.L.R. to Commissioner Charles Edwards (March 31, 1970), records-FDA.

86. Dr. B.H.B. to Commissioner Charles Edwards (March 28, 1970), records-FDA.

87. Dr. J.H.R. to Commissioner Charles Edwards (April 28, 1970), records-FDA.

88. Dr. A.P.R. to Commissioner Charles Edwards (March 27, 1970), records-FDA.

89. "The Pros and Cons of the Pill," *Time,* 93 (May 2, 1969), 58.

90. C. Joseph Stetler, President, Pharmaceutical Manufacturers Association to Hearing Clerk (May 8, 1970), 2, records-FDA.

91. The five symptoms listed in the proposed label were "severe headache, blurred vision, pain in the legs, pain in the chest or unexplained cough, irregular or missed periods." *Federal Register* 35 (April 10, 1970): 5962.

92. Richard D. Lyons, "Official Says Millions Get Pill Illegally," *New York Times,* June 10, 1970, 22.

93. *Federal Register* 35 (June 11, 1970): 9003.

94. Ibid., 9002.

95. "Suit to Seek Access to Birth Pill Data," *New York Times,* June 29, 1970, 61.

96. Ibid.

97. "Birth Control Pills Now Carry Warning on Their Side Effects," *New York Times,* September 10, 1970, 25.

98. Seaman, interviewed by the author, New York, N.Y., January 30, 1994.

99. Edwards had little success in elevating the status or improving the condition of the FDA. The agency continued in disarray well after his tenure was completed in 1973. The *New York Times* reported in 1977 that the FDA was "the Federal Government's most criticized, demoralized and fractionalized agency." Richard Lyons, "Demoralized F.D.A. Struggles to Cope," *New York Times,* March 14, 1977, 1, 49.

100. "History of FDA Patient Package Insert Requirements," *American Journal of Hospital Pharmacy* 37 (December 1980): 1660.

101. F. Gilbert McMahon, "The Patient Package Insert," *Journal of the American Medical Association* 233 (September 8, 1975): 1089.

102. For the establishment of professional autonomy in the medical profession and for subsequent assaults on the autonomy of doctors, see Paul Starr, *The Social Transformation of American Medicine* (New York: Basic Books, 1982).

103. David E. Booker, Irwin R. Pahl, and Donald A. Forbes, "Control of Postpartum Breast Engorgement with Oral Contraceptives, II," *American Journal of Obstetrics and Gynecology* 108 (September 1970): 242.

104. The scope of this section is limited to the involvement of Barbara Seaman and Alice Wolfson—two of the women most active in the pill controversy—in the origins of the women's health movement. Many other organizers and early participants in the movement came from different areas of interest and activism in women's health issues. For a more complete history of the women's health movement, see Sheryl Burt Ruzek, *The Women's Health Movement: Feminist Alternatives to Medical Control* (New York: Praeger, 1978).

105. Ruzek, *Women's Health Movement,* 9.

106. Seaman, interviewed by the author, New York, N.Y., January 30, 1994.

107. For the conflict inherent in La Leche League ideology, which both "promoted women's autonomy and restricted women's roles," see Lynn Y. Weiner, "Reconstructing Motherhood: The La Leche League in Postwar America," *Journal of American History* (March 1994): 1357–81.

108. Janet Carlisle Bogdan, "Childbirth in America, 1650–1990," in *Women, Health, and Medicine in America: A Historical Handbook,* ed. Rima D. Apple (New Brunswick, N.J.: Rutgers University Press, 1990), 119.

109. Barbara Seaman, telephone conversation with author, August 9, 1993.

110. Wolfson, telephone interview by the author, October 25, 1994.

111. Seaman, interviewed by the author, New York, N.Y., January 30, 1994.

112. Ibid.

113. Wolfson, telephone interview by the author, October 25, 1994. Seaman, interviewed by the author, New York, N.Y., January 30, 1994. Ruzek, *Women's Health Movement,* 155; Gena Corea, *The Hidden Malpractice* (New York: Harper Colophon, 1985), 297. The other founders of the National Women's Health Network were Belita Cowan, Mary Howell, and Phyllis Chesler.

114. Ruzek, *Women's Health Movement,* 2.

Chapter Six: Conclusion

1. Deborah Anne Dawson, "Trends in Use of Oral Contraceptives—Data from the 1987 National Health Interview," *Family Planning Perspectives* 22 (July/August 1990): 169.

2. Christine A. Bachrach, "Contraceptive Practice among American Women, 1973–1982," *Family Planning Perspectives* 16 (November/December 1984): 256–57.

3. William D. Mosher, "Contraceptive Practice in the United States, 1982–1988," *Family Planning Perspectives* 22 (September/October 1990): 200, 201. Still, the pill remained the most popular *temporary* contraceptive, used by 21 percent of currently married women, 25 percent of formerly married women, and 59 percent of never-married women using contraception.

4. Elise F. Jones, James R. Beniger, and Charles F. Westoff, "Pill and IUD Discontinuation in the United States, 1970–1975: The Influence of the Media," *Family Planning Perspectives* 12 (November/December 1980): 293–300.

5. Susan Gilbert, "No Link is Found between Pill and Cancer," *New York Times,* September 25, 1996, Sec. C, p. 4. See also the entire September 1996 issue of *Contraception.*

6. H. Tristram Englehardt, Jr. and Arthur L. Caplan, "Patterns of Controversy and Closure: the Interplay of Knowledge, Values, and Political Forces," in *Scientific Controversies: Case Studies in the Resolution and Closure of Disputes in Science and Technology,* eds. H. Tristram Englehardt, Jr. and Arthur L. Caplan (Cambridge: Cambridge University Press, 1987), 11, 14.

7. At the thirtieth anniversary of the pill, experts still vacillated on the relative risks and benefits of oral contraception, demonstrating the intractable problem of resolving large-scale epidemiological studies with individual risk assessment in decisions on birth control. See Frances A. Althaus and Lisa Kaeser, "At Pill's 30th Birthday, Breast Cancer Question is Unresolved," *Family Planning Perspectives* 22 (July/August 1990), 173–76.

8. "Informed Consent Regarding 'The Pill' and New Drugs," *California Medicine* 111 (August 1969), 140.

9. "AMAgrams," *Journal of the American Medical Association* 212 (May 1970), 959.

10. See, for example, Gena Corea, *The Hidden Malpractice: How American Medicine Mistreats Women* (New York: Harper Colophon, 1985). Corea lists dozens of women's health organizations, projects, clinics, and publications, most of which sprang up in the 1970s.

11. Linda Gordon, *Woman's Body, Woman's Right* (New York: Penguin Books, 1990), 386–96.

12. The focus of this study has been restricted to the United States. However, the main story of population control took place overseas. Through the 1970s, when relatively fewer American women used either the pill or the IUD, population control programs aggressively promoted the use of both methods in underdeveloped nations.

13. For a recent articulation of these goals, see Ellen Chesler, "Stop Coercing Women," *New York Times Magazine,* February 6, 1994, 31.

14. Stephanie Coontz, *The Way We Never Were: American Families and the Nostalgia Trap* (New York: Basic Books, 1992), 197–98.

15. Ira Robinson et al., "Twenty Years of the Sexual Revolution, 1965–1985: An Update," *Journal of Marriage and the Family* 53 (February 1991), 217.

16. The abortion pill RU-486 has not been considered here because the issues and controversy surrounding its use pertain to the debate over abortion, not contraception.

17. Matthew Rees, "Shot in the Arm," *New Republic,* December 9, 1991, 16–17; Barbara Kantrowitz and Pat Wingert, "The Norplant Debate," *Newsweek,* February 15, 1993, 37–41.

18. Michael Klitsch, "Injectable Hormones and Regulatory Controversy: An End to the Long-Running Story?" *Family Planning Perspectives* 25 (January/February 1993), 37–40.

19. "United States Loses Lead in Contraceptive Choices, R & D," *Journal of the American Medical Association* 268 (July 8, 1992), 176; Carl Djerassi, "The Bitter Pill," *Science* 245 (July 28, 1989), 360.

20. Natalie Angier, "Future of the Pill May Lie Just Over the Counter," *New York Times,* August 9, 1993, Sec. E, 5; James Trussell et al., "Should Oral Contraceptives Be Available without Prescription?" *American Journal of Public Health* 83 (August 1993): 1094–99.

This essay reviews the main archival sources, books, and journal articles used to research this book. For articles in newspapers, magazines, and medical journals, the reader is directed to specific references in the endnotes.

The records of the Planned Parenthood Federation of America (Sophia Smith Collection, Smith College, Northampton, Mass.), the Population Council (Rockefeller Archive Center, North Tarrytown, N.Y.), the Food and Drug Administration (Historian's Office, Rockville, Md.), and the Worcester Foundation for Experimental Biology (Shrewsbury, Mass.) provided a wealth of information pertaining to the pill, going far beyond the roles of these institutions in the story. Research at the American College of Obstetricians and Gynecologists (Washington, D.C.), the Boston Women's Health Book Collective (Somerville, Mass.), and the Senate Records (National Archives, Washington, D.C.) also uncovered interesting documents. Attorneys at the G. D. Searle & Company (Skokie, Ill.) prevented its Historical Archive from releasing any material not previously published; information about Searle's role in the development, advertising, and marketing of Enovid had to be gleaned from other sources. The Television News Archive at Vanderbilt University (Nashville, Tenn.) provided videotape of televised evening news reports on the Nelson hearings and other related reports on the pill. This archive is a wonderful resource for anyone studying American history after 1968 (when Vanderbilt University began to tape nightly newscasts).

The papers of the following key individuals helped to elucidate their contributions to the development of oral contraceptives and the subsequent safety controversy: Alan F. Guttmacher (Francis A. Countway Library of Medicine, Harvard University, Boston, Mass.), Gregory Pincus (Library of Congress), John Rock (Francis A. Countway Library of Medicine, Harvard University, Boston, Mass.), John D. Rockefeller III (Rockefeller Archive Center, North Tarrytown, N.Y.), and Barbara Seaman (Schlesinger Library, Radcliffe College, Cambridge, Mass.). Transcripts of the 1970 Senate hearings on the pill, along with many supporting documents submitted as evidence to the committee, can be found in volumes 15–17 of the U.S. Senate Select Committee on Small Business (91st Cong., 2nd sess.), *Hearings on Competitive Problems in the Drug Industry* (Washington, D.C., Government Printing Office, 1970).

I learned an enormous amount from my interviews with five individuals involved in the debate over the safety of the oral contraceptives: Ben Gordon, staff economist of the U.S. Senate Small Business Committee during the 1970 Senate hearings on the pill; Morton Mintz, *Washington Post* reporter and author of *The Pill: An Alarming Report;* Judy Norsigian, member of the Boston Women's Health Book Collective and co-author of *Our Bodies, Ourselves;* Barbara Seaman, author of *The Doctor's Case against the Pill;* and Alice Wolfson, member of the feminist group, D.C. Women's Liberation, which demonstrated at the 1970 hearings on the pill.

The Family Planning Oral History Project at Schlesinger Library (Radcliffe College, Cambridge, Mass.) contains transcripts of interviews with important figures in the history of the birth control movement in America. I found the following oral histories to be particularly relevant for my research: Mary Steichen Calderone, M.D., medical director of the Planned Parenthood Federation of America, 1953–64; Mrs. Alan F. Guttmacher, wife of the national director of the Planned Parenthood Federation of America; Grant Sanger, M.D., son of Margaret Sanger; Adaline Pendleton Satterthwaite, M.D., physician at Ryder Memorial Hospital, Humacao, Puerto Rico, during the pill trials; and Christopher Tietze, M.D., and Sarah Lewit Tietze, director and research associate at the National Committee on Maternal Health.

Much of the historical literature on oral contraceptives consists of retrospective accounts or reviews by participants in the development and testing of the pill. These include Frank B. Colton, "Steroids and 'the Pill': Early Steroid Research at Searle," *Steroids* 57 (December 1992): 624–30; Egon Diczfalusy, "Gregory Pincus and Steroidal Contraception: A New Departure in the History of Mankind," *Journal of Steroid Biochemistry* 11 (1979): 3–11; Victor A. Drill, "History of the First Oral Contraceptive," *Journal of Toxicology and Environmental Health* 3 (1977): 133–38; Joseph W. Goldzieher and Harry W. Rudel, "How the Oral Contraceptives Came to be Developed," *Journal of the American Medical Association* 230 (October 1974): 421–25; Frederick S. Jaffe, "Knowledge, Perception, and Change: Notes on a Fragment of Social History," *Mount Sinai Journal of Medicine* 42 (1975): 286–99; David A. Grimes, "The Safety of Oral Contraceptives: Epidemiological Insights into the First 30 Years," *American Journal of Obstetrics and Gynecology* 166 (June 1992): 1950–54; A. D. G. Gunn, *Oral Contraception in Perspective: Thirty Years of Clinical Experience with the Pill* (Park Ridge, N.J.: Parthenon, 1987); Gregory Pincus, *The Control of Fertility* (New York: Academic Press, 1965); Mary Jean Pramik,

Norethindrone: The First Three Decades (Palo Alto, Calif., Syntex Laboratories, 1978); and several books and articles by Carl Djerassi: *The Politics of Contraception* (New York: W. W. Norton, 1979); *The Pill, Pygmy Chimps, and Degas' Horse* (New York: Basic Books, 1992); "The Making of the Pill," *Science* 84, November 1984, 127–29; "The Bitter Pill," *Science* 245 (July 28, 1989): 356–61; and "Steroid Research at Syntex: 'the Pill' and Cortisone," *Steroids* 57 (December 1992): 631–41.

Readers interested in the scientific literature on the pill in the 1960s should begin with the reports and bibliographies in the two FDA publications (Advisory Committee on Obstetrics and Gynecology, Food and Drug Administration, *Report on the Oral Contraceptives*, 1966, and *Second Report on the Oral Contraceptives*, 1969, both published by the U.S. Government Printing Office) and in the volume of papers from the 1969 conference on metabolic effects of the pill, *Metabolic Effects of Gonadal Hormones and Contraceptive Steroids*, eds. Hilton A. Salhalnick, David M. Kipnis, and Raymond L. Vande Wiele (New York: Plenum Press, 1969). For the history of the reproductive sciences, see Merriley Borell, "Biologists and the Promotion of Birth Control Research, 1918–1938," *Journal of the History of Biology* 20 (Spring 1987): 51–87; Adele Clarke, "Controversy and the Development of Reproductive Sciences," *Social Problems* 37 (February 1990): 18–37; Adele Clarke, "Embryology and the Rise of American Reproductive Sciences," in *The Expansion of American Biology*, eds. Keith R. Benson, Jane Maienschein, and Ronald Rainger (New Brunswick, N.J.: Rutgers University Press, 1991); and Nelly Oudshoorn, *Beyond the Natural Body: An Archeology of Sex Hormones* (London: Routledge, 1994).

Several books about the pill that appeared from 1968 to 1970 should be considered primary source material rather than secondary historical research; they include Barbara Seaman's *The Doctors' Case against the Pill* (New York: Peter H. Wyden, 1969), Morton Mintz's *The Pill: An Alarming Report* (Boston: Beacon Press, 1970), Robert W. Kistner's *The Pill: Facts and Fallacies about Today's Oral Contraceptives* (New York: Delacorte Press, 1968), and Jules Saltman's *The Pill: It's Effects, It's Dangers, It's Future* (New York: Grosset and Dunlap, 1970). Paul Vaughan's *The Pill on Trial* (New York: Coward-McCann, 1970) gives an interesting contemporary account of the first decade of the pill, but unfortunately provides no notes on source material. Two articles from the late 1970s offer brief introductions to the history of the development of the pill: Kenneth S. Davis's "The Story of the Pill," *American Heritage*, August–September 1978, 80–91, is directed toward a general audience, while R. Christian Johnson's more nuanced article, "Feminism, Philanthropy and Science in the Development of the Oral Contraceptive Pill," *Pharmacy in History* 19 (1977): 63–78, examines the motivations of the pill's champions. John Rock is the subject of Loretta McLaughlin's interesting biography, *The Pill, John Rock, and the Church* (Boston: Little, Brown, 1982), but unfortunately no sources are listed. John Rock's own views on the birth control pill can be found in his book, *The Time Has Come: A Catholic Doctor's Proposals to End the Battle over Birth Control* (New York: Alfred A. Knopf, 1963).

James Reed chronicles the development of the oral contraceptive in the final section of his fine intellectual history, *The Birth Control Movement and American Society: From Private Vice to Public Virtue* (Princeton: Princeton University Press, 1984). Linda Gordon's outstanding social history, *Woman's Body, Woman's Right: Birth Control in America* (New York: Penguin Books, 1990) ends before the development of the pill. Together these two books provide an excellent introduction to the study of the history of birth control in the United States and should serve as the starting point for all related research. For an analysis of the methodological and ideological differences between these two works, see the review essay by Elizabeth Fee and Michael Wallace, "The History and Politics of Birth Control: A Review Essay," *Feminist Studies* 5 (Spring 1979): 201–15.

Some recent contributions focus exclusively on the pill. Renee Courey's dissertation, "Participants in the Development, Marketing and Safety Evaluation of the Oral Contraceptive, 1950–1965: Mythic Dimensions of a Scientific Solution" (Ph.D. diss., University of California at Berkeley, 1994), is based almost entirely on a single archival source, the Gregory Pincus papers at the Library of Congress. She focuses on the scientific community, categorizing scientists as either proponents or opponents of the development of the pill in the context of the population control movement. Bernard Asbell's *The Pill: A Biography of the Drug that Changed the World* (New York: Random House, 1995) relies heavily on secondary sources; however, his book does offer a nice review of the Catholic response to the pill. On this subject, readers should also consult John T. Noonan's exhaustive study, *Contraception: A History of Its Treatment by the Catholic Theologians and Canonists* (Cambridge: Harvard University Press, 1965) and William H. Shannon's *The Lively Debate: Response to Humanae Vitae* (New York: Sheed and Ward, 1970).

Patricia Gossel uses artifacts from the collections of the Smithsonian's Museum of American History to explore the medical and social ramifications of the "Dialpak" package for birth control pills in "Packaging the Pill," a paper presented at the Workshop on Museums and the History of Technology at the Science Museum of London (July 30, 1996). Lara Marks writes about the early history of the pill in two articles: "'A Cage of Ovulating Females': The History of the Early Oral Contraceptive Pill Clinical Trials, 1950–1959," in *Molecularising Biology and Medicine*, 1930s–1970s, edited by H. Kamminga and S. De Chadarevian (Reading, U.K.: Harwood Academic Press, 1997) and "'Andromeda Freed from her Chains':

Attitudes towards Women and the Oral Contraceptive Pill, 1950–1970" in *Women and Modern Medicine*, eds. Anne Hardy and Lawrence Conrad (Rodopi (Clio-Medica), 1998). Marks's article, "'Not just a statistic': USA and UK policy over thrombotic disease and the oral contraceptive pill, 1960s–1970s" *Social Science and Medicine* (forthcoming 1998), locates the responses of British and American medical experts and government policy-makers to the link between the pill and blood-clotting disorders within the different legal, medical, social, and political traditions of the two countries.

In addition to Reed's and Gordon's studies of birth control in America in the nineteenth and twentieth centuries, several other works focus on topics within this broad framework. General works include Carole R. McCann's *Birth Control Politics in the United States, 1916–1945* (Ithaca, N.Y.: Cornell University Press, 1994); *Birth Control and Controlling Birth: Woman-centered Perspectives*, eds. Joyce Avrech Berkman, Helen B. Holmes, and Michael Gross (Clifton, N.J.: Humana Press, 1980); and C. Thomas Dienes' *Law, Politics, and Birth Control* (Urbana: University of Illinois Press, 1972). The IUD is the subject of Nicole Grant's *The Selling of Contraception: The Dalkon Shield Case, Sexuality, and Women's Autonomy* (Columbus: Ohio State University Press, 1992).

In her article "'Simple Methods' and 'Determined Contraceptors': The Statistical Evaluation of Fertility Control, 1957–1968," *Bulletin of the History of Medicine* 70 (1996): 266–95, Marcia L. Meldrum looks at field trials of nonhormonal contraceptives, namely, IUDs and the "simple methods" of spermicidal creams, jellies, and foams. Although Annette B. Ramirez and Conrad Seipp concentrate on Puerto Rico in their book, *Colonialism, Catholicism, and Contraception: A History of Birth Control in Puerto Rico* (Chapel Hill: University of North Carolina Press, 1983), they do cover an important chapter in the history of the pill, namely, the American-sponsored clinical trials of the pill on that island in the 1950s. Norman Himes' *Medical History of Contraception* (New York: Schocken Books, 1970) is a classic reference text on birth control, but does not cover the period after World War II. *A History of Contraception: From Antiquity to the Present Day* (Oxford: Basil Blackwell, 1990), by Angus McLaren, presents a sweeping history, most useful for its chronicle of the earliest methods of birth control.

Of course, no review of birth control literature can be complete without reference to Margaret Sanger. Her autobiography, while certainly biased in its presentation, still offers a powerful portrait of the birth control advocate and her mission. Of the several biographies written about Sanger, I found Ellen Chesler's *Woman of Valor: Margaret Sanger and the Birth Control Movement in America* (New York: Simon and Schuster, 1992) to be the most insightful. There is no single history of the Planned Parenthood Federation of America. Jon Knowles's *75*

Years of Family Planning in America: A Chronology of Major Events (New York: Planned Parenthood Federation, 1991) lists highlights from the organization's history. Linda Gordon's *Woman's Body, Woman's Right* is the best existing source for the history of Planned Parenthood. Chesler also provides some history of the organization as it relates to the life of Margaret Sanger.

Linda Gordon's *Woman's Body, Woman's Right* provides an excellent analysis of the relationship between the birth control movement and the population control movement in the twentieth century. Other good introductions to the population control movement include Kurt W. Back, *Family Planning and Population Control: The Challenges of a Successful Movement* (Boston: Twayne Publishers, 1989) and Oscar Harkavy, *Curbing Population Growth: An Insider's Perspective on the Population Movement* (New York: Plenum Press, 1995). Nancy Aries reviews the history of federal birth control programs in her article, "Fragmentation and Reproductive Freedom: Federally Subsidized Family Planning Services, 1960–1980," *American Journal of Public Health* 77 (November 1987): 1465–71. William L. Davis looks at legislation pertaining to birth control in "Family Planning Services: A History of U. S. Federal Legislation," *Journal of Family History* 16 (1991): 381–400. For an insider's contemporary perspective on population control in the 1960s, see Phyllis Tilson Piotrow, *World Population Crisis: The United States Response* (New York: Praeger Publishers, 1973). For a history of the Population Council, see Elaine Moss, *The Population Council: A Chronicle of the First Twenty-five Years, 1952–1977* (New York: The Population Council, 1977). Although it is uncritical in its approach, this institutional history contains useful information on the Council's origins and activities. Lee Rainwater presents an early class-based analysis of fertility and family planning in *And the Poor Get Children: Sex, Contraception, and Family Planning in the Working Class* (Chicago: Quadrangle Books, 1960). The essays in *Fertility and Family Planning: A World View*, eds. S. J. Behrman, Leslie Corsa, Jr., and Ronald Freedman (Ann Arbor: University of Michigan Press, 1969) offer further insight into the perceived relationship between contraception and population control in the 1960s. See also Martha C. Ward, *Poor Women, Powerful Men: America's Great Experiment in Family Planning* (Boulder, Colo.: Westview Press, 1986). Much of the data on contraceptive practices in the 1950s and 1960s come from the four volumes of the Growth of American Families studies and the National Fertility studies: Ronald Freedman, Pascal K. Whelpton, and Arthur A. Campbell, *Family Planning, Sterility, and Population Growth* (New York: McGraw-Hill, 1959); Pascal K. Whelpton, Arthur A. Campbell, and John E. Patterson, *Fertility and Family Planning in the United States* (Princeton: Princeton University Press, 1966); Norman B. Ryder and Charles F. Westoff, *Reproduction in the United States, 1965* (Princeton: Princeton

University Press, 1971); and Charles F. Westoff and Norman B. Ryder, *The Contraceptive Revolution* (Princeton: Princeton University Press, 1977). Charles Westoff and Leslie Aldridge Westoff also published a popular version of the 1965 National Fertility Study, *From Now to Zero: Fertility, Contraception and Abortion in America* (Boston: Little, Brown, 1971).

The media played a central role in the dissemination of information about the oral contraceptive to the general public. Several works address the subject of science and the press: *Scientists and Journalists: Reporting Science as News*, eds. Sharon M. Friedman, Sharon Dunwoody, and Carol Friedman Rogers (New York: Free Press, 1986); Hillier Krieghbaum, *Science and the Mass Media* (New York: New York University Press, 1967); Marcel C. LaFollette, *Making Science Our Own: Public Images of Science, 1910–1955* (Chicago: University of Chicago Press, 1990); Dorothy Nelkin, *Selling Science: How the Press Covers Science and Technology* (New York: W. H. Freeman, 1987); and Terra Ziporyn, *Disease in the Popular American Press: The Case of Diphtheria, Typhoid Fever, and Syphilis, 1870–1920* (New York: Greenwood Press, 1988). Two articles deal specifically with the pill and the press: D. A. Grimes, "Breast Cancer, the Pill, and the Press," in *Oral Contraceptives and Breast Cancer*, ed. R. D. Mann (Park Ridge, N.J.: Parthenon Publishing Group, 1990), pp. 309–322; and Elise F. Jones, James R. Beniger, and Charles F. Westoff, "Pill and IUD Discontinuation in the United States, 1970–1975: The Influence of the Media," *Family Planning Perspectives* 12 (November / December 1980): 293–300.

As part of the conceptualization of the pill as a form of medical technology, I delved into topics in the history of technology. The field has come a long way since 1968, when the Illinois Institute of Technology Research Institute in Chicago published the preliminary results of its National Science Foundation-commissioned study of the "retrospective tracing of key events" leading to five important technological innovations, one of which was the oral contraceptive pill (*Technology in Retrospect and Critical Events in Science*). The final report was published five years later as *Interactions of Science and Technology in the Innovative Process: Some Case Studies* (Columbus, Ohio: Battelle, 1973). These studies presented a positivist chronology of scientific findings that with the benefit of hindsight could be linked to the eventual development of the birth control pill. I found Allan Mazur's *The Dynamics of Technical Controversy* (Washington, D.C.: Communications Press, 1981) to be particularly illuminating in the analysis of controversies about the applications of science. Another interesting book on this topic is a collection of essays edited by H. Tristram Englehardt, Jr. and Arthur L. Caplan, *Scientific Controversies: Case Studies in the Resolution and Closure of Disputes in Science and Technology* (Cambridge: Cambridge University Press, 1987). Judith A. McGaw's essay review, "Women and the History of American Technology," *Signs* 7 (Summer 1982): 798–828, gives a very good introduction to the intersection of women's history and the history of technology. This subject is further explored in two works: *Reconstructing Babylon: Essays on Women and Technology* (Bloomington: Indiana University Press, 1991) ed. Patricia H. Hynes and Judy Wajcman, *Feminism Confronts Technology* (University Park: Pennsylvania State University Press, 1991).

The body of literature on women's history is vast; a complete bibliography would fill a volume in itself. I found the following works to be most helpful in providing a context for understanding the history of women in the postwar years: William H. Chafe, *The American Women: Her Changing Social, Economic, and Political Roles, 1920–1970* (New York: Oxford University Press, 1972); Stephanie Coontz, *The Way We Never Were: American Families and the Nostalgia Trap* (New York: Basic Books, 1992); Ruth Schwartz Cowan, *More Work for Mother: The Ironies of Household Technology from the Open Hearth to the Microwave* (New York: Basic Books, 1983); Peter G. Filene, *Him/Her/Self: Sex Roles in Modern America* (Baltimore: Johns Hopkins University Press, 1986); Betty Friedan, *The Feminine Mystique* (New York: Dell, 1974); Rochelle Gatlin, *American Women Since 1945* (Jackson: University Press of Mississippi, 1987); Brett Harvey, *The Fifties: A Woman's Oral History* (New York: Harper Perennial, 1993); Blanche Linden-Ward and Carol Hurd Green, *American Women in the 1960s: Changing the Future* (New York: Twayne, 1993); Rosalind Rosenberg, *Divided Lives: American Women in the Twentieth Century* (New York: Hill and Wang, 1992); and Sheila Rothman, *Woman's Proper Place: A History of Changing Ideals and Practices, 1870 to the Present* (New York: Basic Books, 1978).

Topics in the history of medicine and health care have received a great deal of attention in recent years from historians of women. Two excellent anthologies offer essays on a wide variety of topics and include extensive bibliographies: *Women, Health, and Medicine in America: A Historical Handbook,* ed. Rima D. Apple (New Brunswick, N.J.: Rutgers University Press, 1992) and *Women and Health in America,* ed. Judith Walzer Leavitt (Madison: University of Wisconsin Press, 1984). I especially recommend Adele Clarke, "Women's Health: Life-Cycle Issues," in *Women, Health, and Medicine in America: A Historical Handbook* as an overall introduction to the study of women's health. Emily Martin's anthropological study of American women and their bodies explores the social and cultural constructions of reproduction in *The Woman in the Body: A Cultural Analysis of Reproduction* (Boston: Beacon Press, 1987). Sheryl Burt Ruzek, *The Women's Health Movement: Feminist Alternatives to Medical Control* (New York: Praeger, 1978) provides the best overview of the women's health movement.

There is a rich and varied literature on the treatment of women by the medical profession. A good place to start is Barbara Ehrenreich and Deidre English, *For Her Own Good: 150 Years of the Experts' Advice to Women* (Garden City, N.Y.: Anchor Press/Doubleday, 1978). See also Gena Corea, *The Hidden Malpractice: How American Medicine Treats Women as Patients and Professionals* (New York: William Morrow, 1977); Sue Fisher, *In the Patient's Best Interest: Women and the Politics of Medical Decisions* (New Brunswick, N.J.: Rutgers University Press, 1986); Catherine Kohler Riesmann, "Women and Medicalization: A New Perspective," *Social Policy* 14 (1983): 3–18; Diana Scully, *Men Who Control Women's Health: The Miseducation of Obstetrician-Gynecologists* (New York: Teachers College Press, 1994); and Alexandra Dundas Todd, *Intimate Adversaries: Cultural Conflict between Doctors and Women Patients* (Philadelphia: University of Pennsylvania Press, 1989). Ludmilla Jordanova, *Sexual Visions: Images of Gender in Science and Medicine Between the Eighteenth and Twentieth Centuries* (Madison: University of Wisconsin Press, 1989) and The Brighton Women & Science Group, *Alice Through the Microscope: The Power of Science over Women's Lives* (London: Virago, 1980) look at women as subjects in both medicine and science. The relationship between women and the pharmaceutical industry is the subject of *Adverse Effects: Women and the Pharmaceutical Industry*, ed. Kathleen McDonnell (Toronto: Women's Educational Press, 1986).

I found the following books and articles on more specific topics within the history of women's health care to be helpful in my research. The use and misuse of sex hormones are treated in Barbara Seaman and Gideon Seaman, *Women and the Crisis in Sex Hormones* (New York: Bantam Books, 1977). Susan C. M. Scrimshaw's article, "Women and the Pill: From Panacea to Catalyst," *Family Planning Perspectives* 13 (November/December 1981): 254–62, interprets women's changing attitudes toward the pill. The DES story is told in Diana B. Dutton, *Worse than the Disease: Pitfalls of Medical Progress* (Cambridge: Cambridge University Press, 1988). Judith Walzer Leavitt's *Brought to Bed: Childbearing in America, 1750 to 1950* (New York: Oxford University Press, 1986) stands out among the many books and articles written on the history of childbirth. Lynn Y. Weiner offers a fascinating analysis of breastfeeding and the La Leche League in "Reconstructing Motherhood: The La Leche League in Postwar America," *Journal of American History* (March 1994): 1357–81.

Several books incorporate the subject of birth control into the larger context of reproductive technologies and reproductive choice. These include *Test-Tube Women: What Future for Motherhood?*, eds. Rita Arditti, Renate Duelli Klein, and Shelley Minden (London: Pandora Press, 1984); Gena Corea, *The Mother Machine: Reproductive Technologies from Artificial Insemination to Artificial Wombs*

(New York: Harper and Row, 1985); and Thomas N. Shapiro, *Population Control Politics: Women, Sterilization, and Reproductive Choice* (Philadelphia: Temple University Press, 1985).

From the vast literature on abortion, I consulted Kristin Luker, *Taking Chances: Abortion and the Decision Not to Contracept* (Berkeley: University of California Press, 1975); Kristin Luker, *Abortion and the Politics of Motherhood* (Berkeley: University of California Press, 1984); and Rosalind Pollack Petchesky, *Abortion and Woman's Choice: The State, Sexuality, and Reproductive Freedom* (New York: Longman, 1984). Finally, every student of the history of women and health care should read *Our Bodies, Ourselves,* by the Boston Women's Health Book Collective (New York: Simon and Schuster, 1992).

More general works on the history of health care provided the context for locating the pill within the larger history of medicine in America. These include James H. Cassedy, *Medicine in America: A Short History* (Baltimore: Johns Hopkins University Press, 1991); *The Laboratory Revolution in Medicine* eds. Andrew Cunningham and Perry Williams (Cambridge: Cambridge University Press, 1992); John Duffy, *From Humors to Medical Science: A History of American Medicine* (Chicago: University of Illinois Press, 1993); David J. Rothman, *Strangers at the Bedside: A History of How Law and Bioethics Transformed Medical Decision Making* (New York: Basic Books, 1991); Edward Shorter, *Bedside Manners: The Troubled History of Doctors and Patients* (New York: Simon and Schuster, 1985); Milton Silverman and Philip R. Lee, *Pills, Profits, and Politics* (Berkeley: University of California Press, 1974); Paul Starr, *The Social Transformation of American Medicine* (New York: Basic Books, 1982); and Peter Temin, *Taking Your Medicine: Drug Regulation in the United States* (Cambridge: Harvard University Press, 1980).

I also benefitted from reading the following monographs in the history of medicine that explored the social history of disease: Allan M. Brandt, *No Magic Bullet: A Social History of Venereal Disease in the United States since 1880* (New York: Oxford University Press, 1987); James T. Patterson, *The Dread Disease: Cancer and Modern American Culture* (Cambridge: Harvard University Press, 1987); and Sheila M. Rothman, *Living in the Shadow of Death: Tuberculosis and the Social Experience of Illness in American History* (New York: Basic Books, 1994). For a thorough, well-written investigation of informed consent, see Ruth R. Faden and Tom L. Beauchamp, *A History and Theory of Informed Consent* (New York: Oxford University Press, 1986).

Both contemporary and historical works on sex, sexuality, and the sexual revolution informed my chapter on the pill and the sexual revolution. For books on these topics written in the 1960s, both popular and scholarly, see Helen Gurley Brown, *Sex and the Single Girl* (New York: Pocketbooks, 1962); Helen Gurley

Brown, *Sex and the Office* (New York: Pocketbooks, 1964); Alfred C. Kinsey, Wardell B. Pomeroy, and Clyde E. Martin, *Sexual Behavior in the Human Male* (Philadelphia: W. B. Saunders, 1948); Alfred C. Kinsey, Wardell B. Pomeroy, Clyde E. Martin, and Paul H. Gebhard, *Sexual Behavior in the Human Female* (Philadelphia: W. B. Saunders, 1953); William H. Masters and Virginia C. Johnson, *Human Sexual Response* (Boston: Little, Brown, 1966); Kate Millett, *Sexual Politics* (Garden City, N.Y.: Doubleday, 1970); Vance Packard, *The Sexual Wilderness: The Contemporary Upheaval in Male–Female Relationships* (New York: David McKay, 1968); Ira L. Reiss, *Premarital Sexual Standards in America* (New York: Free Press, 1960); and Ira L. Reiss, *Premarital Sexual Standards* (New York: Sex Information and Education Council of the United States, 1968).

Historical treatments include Beth L. Bailey, *From Front Porch to Back Seat: Courtship in Twentieth Century America* (Baltimore: Johns Hopkins University Press, 1988); John D'Emilio and Estelle B. Freedman, *Intimate Matters: A History of Sexuality in America* (New York: Harper & Row, 1988); Barbara Ehrenreich, Elizabeth Hess, and Gloria Jacobs, *Re-making Love: The Feminization of Sex* (Garden City, N.Y.: Anchor Press/Doubleday, 1986); Linda Grant, *Sexing the Millennium: Women and the Sexual Revolution* (New York: Grove Press, 1994); Ira L. Reiss, *An End to Shame: Shaping Our Next Sexual Revolution* (Buffalo, N.Y.: Prometheus Books, 1990); Paul Robinson, *The Modernization of Sex: Havelock Ellis, Alfred Kinsey, William Masters and Virginia Johnson* (New York: Harper Colophon, 1976); and Daniel Scott Smith, "The Dating of the American Sexual Revolution: Evidence and Interpretation," in *The American Family in Social–Historical Perspective*, ed. Michael Gordon (New York: St. Martin's Press, 1973), 321–35.

I also consulted several books on the 1960s, to get a feel for social movements that were contemporaneous with the contraceptive revolution. David Chalmers' *And the Crooked Places Made Straight: The Struggle for Social Change in the 1960s* (Baltimore: Johns Hopkins University Press, 1991) gives the best general overview of that decade. A small selection of the large body of literature on the sixties includes Jim F. Heath, *Decade of Disillusionment: The Kennedy–Johnson Years* (Bloomington: Indiana University Press, 1975); Godfrey Hodgson, *America in Our Time* (Garden City, N.Y.: Doubleday and Co., 1976); Edward P. Morgan, *The Sixties Experience: Hard Lessons about Modern America* (Philadelphia: Temple University Press, 1991); William L. O'Neill, *Coming Apart: An Informal History of America in the 1960's* (Chicago: Quadrangle Books, 1971); and Milton Viorst, *Fire in the Streets: America in the 1960s* (New York: Simon and Schuster, 1979). As background to the 1960s, David Halberstam's *The Fifties* (New York: Villard Press, 1993) provides an anecdotal history of the previous decade.

ABC News, 110, 112
Abortion, 37, 60, 131; as birth control method, 8; fight for legalization of, 129, 133; illegal, 3, 79, 108, 115, 119
Abortion pill (RU-486), 161
Advertising, medical, 35–40
African Americans, 30, 56, 62
AIDS, 136
American Cancer Society, 43, 82
American College of Obstetricians and Gynecologists, 97, 102, 124, 126
American Medical Association, 13, 81, 83, 153; Council on Drugs, report on the pill by, 87–88; and patient package insert, 124, 126, 134
Andromeda, 37
Anesthesia, 52
Antibiotics, 6, 19, 105, 107–8
Archer, Elaine, 119
Arthritis, 21, 23
Atomic bomb, 19, 66

Birth control: acceptance of, 11–12; advocates of, 1, 19–20; and AIDS, 136; barrier methods of, 3, 20, 38, 76, 101, 135; as controversial, 7; definition of, 4; efficacy of, 11–12, 79; government subsidy of, 67; laws against, 12–13, 24, 31; for males, 20; medicalization of, 2, 38, 52, 98–99, 108, 116, 131, 137; as medical practice, 7, 12–13, 79, 134, 137; military personnel and, 69; in 1950s, 11–13; physicians and, 7, 12–13, 40, 98; and population control, 3, 4, 19, 67–69; re-

search and development on, 136–37; and risk, 3, 79, 88, 92–93; Roman Catholics and, 11, 13, 17, 24, 29, 31, 45–47, 61–63, 146; as social problem, 1, 21, 69, 75, 135; use of, 61–64; as women's responsibility, 20, 27, 131; women's right to, 137. *See also* Abortion; Cervical cap; Condoms; Depo-Provera; Diaphragms; Douches; Infanticide; Intrauterine devices; Norplant; Pessaries; Pill; Rhythm method; Spermicides; Sterilization; Suppositories; Tubal ligation; Vasectomy; Withdrawal
Birth control pill. *See* Pill
Birth defects, 33
Birth rate, 4, 9, 16, 62–63
Blood clots. *See* Thromboembolism
Boston Women's Health Book Collective, 104, 129
Bronk, Detlev, 18
Brown, Helen Gurley, 54
Brown University, 65
Buck, Pearl S., 66

Calderone, Mary S., 40, 49, 81
California Medical Association, 133
Cancer, 31, 43–44, 77, 82–83, 84, 89, 96, 110, 118; breast, 43, 83, 87, 97, 133; cervical, 43, 78, 82, 93–94; and Depo-Provera, 136; FDA report on the pill and, 86–87; ovarian, 133; uterine, 87, 133
Carson, Rachel, 104, 114
Catholics. *See* Roman Catholics

CBS News, 110–11
Cervical cap, 135
Chang, Min-Chueh, 27–28, 85
Charles Pfizer, Inc., 24
Chemical Specialties Company, Ltd., 28
Childbirth, 20, 52, 129, 134
Children: cost of, 69; rearing of, 20. See also Pregnancy
Ciba Pharmaceuticals, 28
Civil rights movement, 53, 59
Class, socioeconomic: and access to birth control, 67; Connell and, 118; eugenics and, 16; and IUD, 70–71; and the pill, 2, 4–5, 56, 70–71, 101, 128, 135; Planned Parenthood and, 114
Clinics, birth control: first, 13; funding for, 68; and the pill, 40–41, 71, 113–14, 116, 124, 137; state-run, 68
Clinton, Bill, 30
Coitus interruptus. See withdrawal
Cold War, 16, 48
College students: in 1950s, 9; and the pill, 44–45, 65, 67; and premarital sex, 57–61, 136
Colton, Frank B., 23–24
Communism, 16
Compton, Karl T., 18
Comstock Act, 14
Condoms, 8, 11, 13, 54, 60, 137; AIDS and, 136; numbers using, 61–62; teenagers and, 64
Connell, Elizabeth, 118–20
Consumer movement, 104, 116, 133
Contraception. See Birth control
Contraceptive, oral. See Pill, the
Contraceptive revolution, 2, 45, 53, 55–56, 60–63, 65, 84
Cortisone, 21, 23, 25
Counterculture, 53, 54, 73
Cronkite, Walter, 110–11

Dalkon Shield, 157n.28
Davis, Hugh J., 110–11, 157
D.C. Women's Liberation, 129–30; alternative hearings on the pill, 119–20; and Guttmacher, 115; and Senate hearings on the pill, 108–12, 118; sit-in at Department of HEW, 122
Defense, Department of, 69
Demographers: contraceptive use studies, 2, 55, 60–64, 100; and overpopulation, 16–17
Depo-Provera, 131, 136
Detailmen, 35–37, 51, 106
Diaphragms, 11, 53, 54, 60, 116, 135; and Melamed study, 93–94; numbers using, 61–62; physicians and, 13, 38, 99; Sanger and, 14
Dick, David, 111
Dick-Read, Grantly, 129
Diethylstilbestrol (DES), 131, 133
Diosgenin, 22–23
Djerassi, Carl, 24
Doctor-patient relationship, 3–4, 37–40, 50–52, 101–2, 104, 109, 111, 113, 115–21, 123–28
Doctors. See Physicians
Doctors' Case against the Pill, The, 101, 103–7
Dole, Bob, 110, 117
Donaldson, Sam, 110
Double standard, 57. See also Morality
Douches, 8, 11, 13, 62
Draper Committee, 67
Drill, Victor A., 89–90
Drug industry. See Pharmaceutical industry
Drug salesmen. See Detailmen

Economics: of childrearing, 2, 69; development in Third World, 16, 68, 71; and overpopulation, 72; and population control, 134–35; and research, 20, 21
Edwards, Charles: and FDA, 127; meeting with D.C. Women's Liberation, 122; and patient package insert, 113, 118, 120, 124, 126; on safety of the pill, 112–13
Eisenhower, Dwight D., 67–68
Eisenstadt v. Baird, 12, 136
Endocrinology, 20–23, 27, 46–47, 75
Endometriosis, 37

Enovid, 25, 42, 49, 62, 85; advertisements for, 37–39; and blood clots, 43, 51, 80–82; and cancer, 43–44, 82–83; estrogen in, 31; FDA approval of, 32–34, 36, 144; field trials of, 30–32; and gynecologic disorders, 36; indications for use, 37; promotional literature on, 34–40

Environmentalists, 117

Epidemiology, 75, 86, 87, 98, 133; British studies, 89–91; methodology in, 77–78, 90; problems in studying the pill, 78; prospective studies, 77–78; retrospective studies, 78; use of prevalence rates, 93–94, 154

Estrogen, 30, 134; and cancer, 29, 87; in Enovid, 31; in hormone replacement, 128; and infertility, 29; in menstrual cycle, 22; in the pill, 22, 128; synthetic, 22

Eugenics, 16, 21, 141

Experimentation: on animals, 28, 30; on humans, 29–30, 109

Fallout, radioactive, 75, 85

Family planning. *See* Birth control; Population control

Feminists: and abortion, 115, 129, 133; and Connell, 118–20; and medical profession, 2, 3, 52, 102, 108, 111, 116–17, 118–20, 129, 131; and Nelson, 108–10, 112; and new contraceptive methods, 136; in 1960s, 2, 4, 111; in 1970s, 134–35; and patient package insert, 126–28; and the pill, 31, 109, 119, 131; radical, 108, 130; reform, 108, 120, 130; and women's health, 76, 104, 108, 120, 129–31, 134; and women's health movement, 128–31

Fertility, 17, 32, 63

Fertilization, 21, 22, 27–28

Field pea extract, 42

Finch, Robert, 122, 130

Food and Drug Administration (FDA): ad hoc committee on thromboembolism, 43, 82; and drug approval, 74, 83; —Depo-Provera, 136; —Enovid, 25, 32–33, 37, 40, 42, 85, 107; —Ortho-Novum, 38; and feminists, 130; and neuro-ophthalmologic disorders, 84; in 1970s, 159; and patient package insert, 4, 7, 105, 118, 120–28; and the pill: —hearings, 106, 109, 112–13; —in 1990s, 137; —reports on, 62, 86–88, 90–91, 93–96, 97, 103, 112; task force on carcinogenic potential, 82; and thalidomide, 33

Food, Drug, and Cosmetic Act, 32–34, 95–96, 120–21, 125

Freedom of Information Act, 123, 126

Friedan, Betty, 10

Gagnon, John, 59

Gallup poll, 47, 100, 115–16, 118. *See also* Public opinion

Garcia, Celso-Ramon, 81

Gordon, Ben, 106–7

Government, 2; and birth control, 67–70; and medical practice, 3, 4, 124–25, 127, 134; 1960s critique of, 73, 123; and pharmaceutical industry, 104, 106, 127; and population control, 16; and Tuskegee study, 30

Griswold v. Connecticut, 12, 68

Growth of American Families studies, 61–64

Gruening, Ernest, 68–69

Guttmacher, Alan, 43, 70–71, 114–16

Haiti, 32

Health: and feminist ideology, 8; and pharmaceutical industry, 106; and the pill, 3; women's, 2, 3, 86, 103–5, 109, 116, 120, 129–31, 134, 137

Health, Education, and Welfare, Department of, 68, 69, 121, 122, 126

Hearings on Competitive Problems in the Drug Industry, 4, 7, 106–20, 121, 123, 129–30, 134; alarm caused by, 110, 113–17; disruptions of, 108–9, 112

Hefner, Hugh, 54

Hertz, Roy, 87

Hormones, 18, 21–24, 47; and cancer, 83; chorionic gonadotropin, 22; as drugs, 23; follicle-stimulating, 22; luteinizing, 22; 19-nor, 24, 28, 30; plant, 22–23, 25; steroid, 20, 22–24, 36, 38, 83; synthesis of, 22–24; synthetic steroidal, 5, 24, 25, 27, 28–29, 36, 39, 42, 46, 83, 136. *See also* Estrogen; Progesterone
Humanae Vitae, 47, 62–63
Huntley, Chet, 111

Immunology, 18, 20
Infanticide, 8
Infertility, 29, 37
Informed consent, 3, 8, 104, 107, 109, 114–20, 123–26, 129–31, 133–34
International Planned Parenthood Federation, 14, 92
Intrauterine devices (IUD), 8, 60, 94, 116, 131, 133, 135; Davis and, 157; development of, 142; numbers using, 62, 151; physicians and, 99; and population control, 69, 70–72

Johnson, Lyndon B., 68, 69
Johnson & Johnson, 24
Johnson-Reed Immigration Act, 16
Journalists. *See* Media, news
Julius Schmid Pharmaceuticals, 38

Kefauver-Harris Amendments, 32–33, 95–96. *See also* Food, Drug, and Cosmetic Act
Kinsey, Alfred, 57, 59, 66

La Leche League, 129
Lawsuits, pill-related, 7, 81, 125, 126, 127, 134
Ley, Herbert, 127
Lithospermum extract, 42

Magazines, popular. *See* Media, news
Males: contraceptives for, 20, 119; as sex partners, 64, 71
Malthus, Thomas, 15–16
Marker, Russell E., 22–23

Marriage, 12, 45, 53–54, 85; age at, 9, 136; critique of, 59; rate, 9
Mastectomy, 129
McCormick, Katherine Dexter, 20, 21, 26–28, 33, 131, 135
Media, news, 2, 73, 100; and cortisone, 23; information about science, technology and medicine, 6–7, 74, 91; and the pill: —controversy over safety of, 76, 123; —early coverage of, 6, 35–36, 41–50; —and FDA report on, 95; —and health effects of, 80, 83, 84–86, 88–89, 91–93, 99–102, 133; —and hearings on, 107, 110–13, 117, 123; —and history of, 5; and sexual revolution, 2, 54–55, 57, 64–67, 72
Medicaid, 106, 137
Medical practice: attitudes toward, 2, 33, 127, 132; critique of, 73, 107; and the pill, 3, 37–40, 43. *See also* Physicians
Medical profession. *See* Physicians
Medicare, 106
Melamed, Myron R., 93–94
Menstrual cycle, 1, 22, 28, 32, 37–38, 42, 47; menstruation, 22, 29, 30, 37; use of the pill to regulate, 46
Merck, Sharpe & Dohme, 83
Mintz, Morton, 85, 92, 117
Mitford, Jessica, 104
Morality, and the pill, 2, 44–45, 47, 55–56, 64–67, 73, 84, 85, 91
Motherhood, 10, 72; as career, 11; "scientific," 129

Nader, Ralph, 104, 127
Nadler, Jerrold, 105
National Academy of Sciences, 17, 18
National Center for Health Statistics, 82
National Committee on Maternal Health, 81
National Fertility Studies, 61–64, 100
National Institutes of Health, 25
National Prescription Audit, 62
National Science Foundation, 25
NBC News, 110–11, 113
Nelson, Gaylord, 104–19, 127

Nelson hearings. See Hearings on Competitive Problems in the Drug Industry
Neuro-ophthalmologic disorders, 80, 83–84, 87
New Left, 53
Newspapers. *See* Media, news
Norethindrone, 24, 25, 28, 31, 38, 143
Norethynodrel, 24, 25, 28, 30, 31, 32, 38
Norinyl, 38
Norlestrin, 38
Norplant, 131, 136
Notestein, Frank, 18
Nuclear energy, 1, 47

Obstetrician-gynecologists, 80, 91, 97–99, 104. *See also* Medical practice; Physicians
Office of Economic Opportunity, 67, 68
Office of Population Research, 18
Oral contraceptive. *See* Pill, the
Ortho-Novum, 38, 48
Ortho Pharmaceutical Company, 24, 38
Osborn, Frederick, 18
Our Bodies, Ourselves, 129
Overpopulation: and birth control, 68–72; "crisis" of, 2; demographers and, 16; and the pill, 47–49, 75–76, 117, 135; Puerto Rico and, 31; Rock and, 29; Sanger and, 15. *See also* Population control
Ovulation, 1, 22, 27–30, 37, 46, 92

Packard, Vance, 58–60
Pap smears, 137
Parke-Davis and Company, 24, 38
Parran, Thomas, 18
Paternalism, 31, 71, 116, 128
Patient package insert, 4, 7, 105, 113, 118, 120–28, 133, 158
Patients' rights movement, 133, 158
Paul VI (pope), 47
Pessaries, 8
Pesticides, 75, 85, 114
Pharmaceutical industry: and chemical research, 23; critique of, 73–75, 104, 123, 131; and FDA, 106, 127; and med-
ical profession, 105–7; and patient package insert, 105, 118, 120–21, 125–26; and the pill: —estimates of use, 62; —hearings on, 105–7, 109–12; —impact of, 2, 5; —marketing of, 35–40, 51, 71, 92; role in development of, 24–25; power of, 7, 123, 127, 131
Pharmaceutical Manufacturers Association, 125–26
Pharmacists, 12, 33, 99, 121
Phosphorylated hesperidin, 42
Physicians: authority of, 4, 7, 52, 98–99, 117, 121, 126–28, 131, 137; and birth control, 7, 12–13, 40, 98; communication among, 79–80; critique of, 73–74, 104, 109, 119–20, 131; and FDA reports on the pill, 95–96; and information, 3, 75, 79–80, 106, 113, 115–16, 119–21, 134; and informed consent, 133–34; and patient package insert, 105, 121, 124–26; and pharmaceutical industry, 105–7; and the pill, 2, 5, 33, 35–40, 49, 76, 96–99; —acceptance of, 1, 7, 34–35, 51, 96–97; —and adverse health effects of, 43, 87–88, 91, 96–99, 101–2; —and controversy over, 75–76, 85, 96–99, 101–2; —prescribing of to unmarried women, 65, 66; and prescription drugs, 120–21; and risk, 79; and Sanger, 14–15, 131; specialization of, 80, 91; surveys of, 12–13, 35–36, 97, 99, 102. *See also* Doctor-patient relationship; Medical practice
Pill, the: adverse health effects of, 1, 3, 4, 7, 39, 43–44, 72, 74, 76, 78–102, 104–7, 111, 118; advertising of, 35–40, 86, 101; animal experiments, 27–28; and blood clots, 80–82, 84, 89–91, 96; and cancer, 82–83, 84, 89, 93–94, 96, 118; clinical trials of, 28–32, 39, 42, 107, 109; commercial success of, 7, 48, 74, 119; cosmetic effects of, 72; cost of, 35, 39, 40, 48, 134, 137; development of, 7, 27–33, 52; efficacy of, 8, 32, 54, 76, 79; FDA approval of, 32–33, 36, 40, 42; field trials of, 30–32, 109; and financial reward

Pill, the (*continued*)
for physicians, 35, 39, 96, 99, 119; funding for, 20, 25–27; as genocidal, 56; hearings on, 4, 7, 106–20, 121, 123, 129–30, 134; historical accounts of, 5, 42; idea for, 5, 14, 26; information about, 3, 6, 39, 41–49, 87, 102, 104, 107, 109, 112–13, 115–16, 122, 127, 134; marriage and, 38, 45, 48, 67; media coverage of, 35, 41–50, 64–67, 76, 84–86, 88–89, 99–102, 107, 110–13, 117, 123, 133; medical reports on, 86–96, 100–102; metabolic effects of, 86–87, 94–96, 107; mode of action, 22, 42; morality and, 2, 44–45, 47, 55–56, 64–67, 73, 84, 85, 91; and neuro-ophthalmologic problems, 80, 83–84, 87; number of women using, 34–35, 41, 61–64, 100–101, 110–11, 113, 115, 132–33, 136, 151, 160; and patient package insert, 4, 7, 105, 113, 118, 120–28, 133, 158; physicians and, 1, 33–40, 96–99; Pincus and, 20–21, 25–33, 50; and population control, 3, 47–49, 67–72, 84, 88, 91, 161; prescription-only status of, 4–5, 35, 38, 50–51, 64, 116, 137; and promiscuity, 45, 65–67, 71; and risk, 7, 75, 78–79, 88, 92–93, 95–96, 100–102, 107; Roman Catholics and, 45–47, 61–63; safety of, 3, 7, 31–33, 38, 76, 79, 85–86, 98, 100–103, 107, 109–13, 132; sale of, 7, 34–40, 51; sexual revolution and, 2, 53–57, 60–61, 64–67, 88; side effects of, 32, 37, 39, 41, 43, 48, 76–77, 85, 87, 101, 118, 122, 126; social impact of, 1, 5, 7–8, 42, 45, 47, 64–67, 71–72, 85, 88–89, 135–37; as symbol of social change, 2, 55, 84; as technology, 6, 71–72, 75, 85, 92, 120, 132; teenagers and, 63–64; as "therapy," 39, 50, 79, 88; use by healthy women, 32, 39, 74, 79, 88, 92, 95; women's acceptance of, 1, 6, 7, 33–35, 49–52, 62, 72
Pincus, Gregory, 81; and development of the pill, 25–33; and media, 42, 85; requests for the pill, 49–50; and Sanger,

20–21; and Searle, 21, 24–25; study of the pill and cancer, 43, 82–83
Piotrow, Phyllis, 117, 118
Planned Parenthood: of Chicago, 113; of Denver, 49; of Detroit, 116; Eisenhower and, 68; and family planning, 69; of Los Angeles, 39, 143; and Melamed study, 93; of New York, 93–94, 116; and patient package insert, 124; and Pincus, 25–26, 28; of Pittsburgh, 56; and the pill: —health effects of, 100–101; —and hearings on, 113–16, 117; —and use of, 34–35, 36, 40–41, 43, 116, 137; —requests for, 49; and Population Council: 18–19; and Seaman, 103
Playboy, 43, 54, 55
Population control: advocates of, 2–3, 67, 70–72, 117, 134–35; and birth control, 3, 4, 19, 47–49, 67–68, 114; definition of, 4; and eugenics, 16, 49, 141; feminist critique of, 134–35; and hearings on, 117; the pill and, 47–49, 67–72, 75–76, 84, 88, 91, 161; Rock and, 29; roots of, 15–19; as social problem, 1; and U.S. foreign policy, 67. *See also* Overpopulation
Population Council, 17–19, 69, 119; and basic research, 18; biomedical division, 18; and IUD, 70, 134–35; and patient package insert, 159; and the pill, 26, 62, 134–35
Population Crisis Committee, 117, 118
Potts, Malcolm, 92
Pregnancy: cost of, 69; as disease, 79; fear of, 54, 60, 136; as female responsibility, 20; and infertility, 29; and the pill, 1, 8, 88, 92, 96; risk of, 3, 79; and thalidomide, 33; unwanted, 2, 12, 50, 66, 79, 113–14, 117, 118–19
Press, popular. *See* Media, news
Privacy, right to, 12
Progesterone, 22–24, 25, 27–31; and cancer, 87; synthetic, 24, 25, 27–31, 83
Progressive Era, 27
Promiscuity, 45, 65–67, 71. *See also* Morality

Pronatalism, 11
Public, general: and the pill, 2, 6, 20, 41–49, 51, 56, 73, 84, 118, 127; —controversy over safety of, 76, 86, 96, 137; —and FDA report on, 95; —and hearings on, 113–16; faith in science and technology, 6, 19–20, 21, 33, 44, 85, 92; understanding of medicine, 116, 118, 126–27
Public health, 16
Public opinion, 44; of the pill, 7, 61, 64, 74, 100; of science, 6; shaped by media, 64; of women's roles, 10. *See also* Gallup poll
Puerto Rico, 31–32, 42, 48

Racism, 56
Religion: and birth control, 45–47; and the pill, 45–47, 62–63, 84, 85, 91. *See also* Roman Catholics
Reproductive physiology, 8, 18, 20, 21–24, 27, 46–47. *See also* Menstrual cycle
Rhythm method, 8, 11; numbers using, 61–62; physicians and, 38; Roman Catholics and, 45–47, 61, 62–63
Rice-Wray, Edris, 31–32, 48
Risk: and contraceptive choice, 3, 74, 113; mentioned in patient package insert, 118, 122; and pill use, 7, 75, 78–79, 88, 92–93, 95–96, 100–102, 107
Rock, John, 28–32, 42, 46–47, 81–82, 85
Rockefeller, John D., III, 17–18
Rockefeller Foundation, 17, 28
Roman Catholics: and birth control, 11, 17, 24, 45–47, 146; as physicians, 13; and pill, 45–47, 61–63; in Puerto Rico, 31; Rock and, 29. *See also* Religion

Sanger, Margaret, 13–15, 119; attitude toward science, medicine, and technology, 15, 18, 131; concern about overpopulation, 15; and first birth control clinic, 13; and McCormick, 20–21, 26–27; and medical profession, 14–15, 131; and the pill, 14, 19, 21, 28, 33, 54, 131; and Pincus, 20–21, 26–27; and population control, 15; and Rock, 29; and

woman-controlled contraception, 14, 20, 54, 131, 135
Sartwell, Philip, 90–91
Science: attitudes toward, 6, 30, 33, 49, 73, 85, 108, 132; McCormick and Sanger and, 27; and the pill, 5, 20, 21, 41, 84; and religion, 46–47; to serve society, 15, 18, 49, 72, 85
Scientists: and controversy, 75, 78; as experts, 27; and hearings on the pill, 117; and physicians, 98; and the pill, 1, 2, 14, 43, 49, 92, 94, 96; and research in reproductive biology and steroid chemistry, 21–24; and Sanger, 14–15, 131
Seaman, Barbara: book on the pill, 101, 103–5; and D.C. Women's Liberation, 108, 119, 122; and hearings on the pill, 105–9, 114; and National Women's Health Network, 109; and patient package insert, 127; reports on the pill, 85–86; and women's health movement, 129–30, 134
Searle, G.D., and Company: and competition from IUDs, 70–71; conference sponsored by, 80–82; and Drill, 90; and Enovid, 31–33, 35–40, 42, 62; and financial success, 48, 51; and hearings on the pill, 111–12; and Pincus, 21, 24–25, 28; and steroid research, 23–24, 31; stock, 44, 51
Sex: attitudes toward, 2, 8, 21, 45, 55, 57–61, 69, 71; extramarital, 54, 65–66; impact of pill on, 45, 85; liberalization of, 2, 53–54, 69; premarital, 2, 12, 45, 56–61, 65–67, 136; public discussion of, 54–55, 59, 65, 69–70
Sexual revolution, 2, 45, 53–61, 64–67, 71–72, 135–36
Side effects: breakthrough bleeding, 31, 39, 43, 77, 87; breast tenderness, 43, 76, 87; dizziness, 32, 43; gastrointestinal discomfort, 32, 39, 43, 76; headache, 32, 43; nausea, 32, 39, 76, 87; weight gain, 39, 43, 77, 87. *See also* Cancer; Neuro-ophthalmologic disorders;

Side effects (*continued*)
Stroke; Thromboembolism; Thrombophlebitis
Smoking, cigarette, 87, 92
Society, American: birth control and, 7, 13; feminist critique of male-dominated, 117, 120, 129; and medical science, 108, 135–36; in 1960s, 2, 53–55, 72, 73, 123; the pill and, 1, 42, 52, 57, 64, 71–72, 135–36; power in, 120; sex and, 55
Sociologists, studies of premarital sex by, 2, 55–61
Spermicides, 11, 14, 62, 137
State, Department of, 69
Steinem, Gloria, 45
Steinfeld, Jesse, 122
Sterilization: abuse, 129; as birth control method, 8, 60, 62, 132–33; and eugenics, 16; laws, 119
Steroid chemistry, 20, 23–24, 75
Steroids. *See* Hormones
Stone, Abraham, 21
Strauss, Lewis, 18
Stroke, 77, 84, 91, 117, 133
Student movement, 53
Suburbs, 9–10
Suppositories, vaginal, 8
Syntex Corporation, 24, 25, 28, 31, 38, 143
Syphilis, 30

Technology: attitudes toward, 2, 132; choice of, 19–21; contraceptive, and fertility control, 63, 71; medical, the pill as, 1, 6, 8, 69, 75–76, 92, 108, 120, 132; misgivings about, 73, 85; NSF report on, 139–40; as solution to social problems, 1, 6, 18, 72, 92, 120, 135
Teenagers, 57, 60–61, 63–64, 67
Television, 6, 55, 65, 85; coverage of hearings on the pill, 110–13
Thalidomide, 33, 73, 85
Thromboembolism, 77, 78, 96; and the pill: —AMA report on, 87; —British studies of, 89–91; —conference on, 80–82; —and Enovid, 38, 43, 80–82;

—FDA report on, 86–87; —and hearings on, 110, 117; —Sartwell study, 90–91; first report of, 80–81; patient package insert mention, 122, 126; and Searle, 51
Thrombophlebitis, 43, 81, 84
Tietze, Christopher, 81, 119
Tietze, Sarah Lewit, 119
Tubal ligation, 116. *See also* Sterilization
Tuberculosis, 52
Tuskegee, 30
Tyler, Edward T., 39, 143

Underdeveloped countries, 4, 16, 17, 67, 70, 135, 161; women in, 134
United Nations, 16
University of Puerto Rico, 29–30, 31
Upjohn Company, 24, 25
U.S. Circuit Court of Appeals, 14
U.S. Congress, 68–69, 105–6, 111, 130
U.S. Public Health Service, 30, 69
U.S. Senate Subcommittee on Monopoly of the Select Committee on Small Business, 3–4, 104–6, 116, 117
U.S. Supreme Court, 12, 68, 136

Vaccine, 88; for birth control, 20; for polio, 6, 30
Vaginal orgasm, myth of, 54, 148
Vasectomy, 116, 133. *See also* Sterilization
Vietnam War, 53, 59, 123
Virginity, 44, 57, 63, 66

Welfare, 56, 119, 136; maternal and child, 16
Winter, Irwin, 111–12
Withdrawal, 8, 11, 62, 64
Wolfson, Alice, 109, 122, 129–30
Women: changing roles of, 57, 63, 71, 135; college-educated, 9, 16, 34, 44–45, 71; contraceptive use, 2, 12, 55, 60–64, 141; demand for information, 3, 102, 119–20, 123; in 1950s, 9–13; as patients, 50–52, 102, 116–17, 119–20; and the pill: —acceptance of, 1, 7, 33, 49–52, 62, 101; —and adverse health effects of, 88,

99–102, 111; —dissatisfaction with, 76, 99, 101–2, 115; —experiences with, 102, 108, 111, 119; —hearings on, 113, 115;—number using, 34–35, 41, 61–64, 100–101, 110–11, 113, 115, 132–33, 136, 151, 160; —requests for, 49–52; —unmarried women and, 2, 34, 44–45, 56, 63–67, 90, 135–36, 160; and risk assessment, 79, 101–2, 113; and sterilization, 132–33; working, 10–11, 71, 135
Women's health movement, 3, 8, 104, 128–31, 133–35, 160
Women's organizations: Bread and

Roses, 130; National Organization for Women, 130; National Women's Health Network, 130; Redstockings, 130. *See also* D.C. Women's Liberation
Worcester Foundation for Experimental Biology, 20–21, 25–29, 49
Worcester State Hospital, 30
World Health Organization, 86–87, 97
World War II, 6, 16, 19, 29, 52, 129
Wyden, Peter, 86, 105
Wynn, Victor, 112

Library of Congress Cataloging-in-Publication Data

Watkins, Elizabeth Siegel.

    On the pill : a social history of oral contraceptives, 1950–1970 /
Elizabeth Siegel Watkins.

       p.   cm.

    Includes bibliographical references and index.

    ISBN 0-8018-5876-3 (alk. paper)

    1. Birth control—United States—Social aspects.   2. Oral
contraceptives—United States—History.   3. Sex customs—
United States—History—20th century.   I. Title.

HQ766.5.U5W325   1998

363.9'6'0973—dc21   98-5003   CIP